Glycerol

The Renewable Platform Chemical

Mario Pagliaro
Consiglio Nazionale delle Ricerche

ELSEVIER
elsevier.com

Elsevier
Radarweg 29, PO Box 211, 1000 AE Amsterdam, Netherlands
The Boulevard, Langford Lane, Kidlington, Oxford OX5 1GB, United Kingdom
50 Hampshire Street, 5th Floor, Cambridge, MA 02139, United States

British Library Cataloguing-in-Publication Data
A catalogue record for this book is available from the British Library

Library of Congress Cataloging-in-Publication Data
A catalog record for this book is available from the Library of Congress

ISBN: 978-0-12-812205-1

Publisher: John Fedor
Acquisition Editor: Katey Birtcher
Editorial Project Manager: Emily Thomson
Production Project Manager: Anitha Sivaraj
Cover Designer: Alan Studholme

Dedication

This book is dedicated *to Francesco Meneguzzo, eminent scientist and friend.*

Contents

About the Author

Mario Pagliaro is a chemistry and energy scholar based at Italy's Research Council in Palermo, Italy, where he leads a research Group focusing on nanochemistry, sustainability, and the bioeconomy. In recognition of his "significant contributions to the chemical sciences" in 2014 he was designed Fellow of the Royal Society of Chemistry.

The achievements of his Group's research developed in co-operation with leading researchers based in 20 countries, including Israel, Canada, China, Portugal, and the United States, include numerous important advances reported in over 180 research papers and several book chapters.

Mario received his doctorate in chemistry from the University of Palermo in 1998, following work carried out with David Avnir at the Hebrew University of Jerusalem, and Arjan de Nooy at the TNO Food Research Institute in Zeist (the Netherlands). In late 2000 he joined the ranks of Italy's Research Council at the Institute of Chemistry and Technology of Natural Products founded by Professor Giulio Deganello.

Frequently cited for his excellence in teaching, in 2005 he was appointed as "Maître de conférences associé" at Montpellier's Ecole Nationale Superieure de Chimie. In 2008 he gave the "John van Geuns" Lecture at the University of Amsterdam. In 2010 he was invited to give a keynote lecture at the eighth Eurofedlipid Congress in Munich. In 2012 he gave an invited lecture at the 22nd Canadian Symposium on Catalysis, and an invited seminar on solar hydrogen at Zurich's ETH.

He regularly organizes conferences and gives courses and seminars on the topics of his research. He co-chairs the "SuNEC Sun New Energy Conference" as well as the "FineCat Symposium on heterogeneous catalysis for fine chemicals"—two important meetings held annually in Sicily since 2011.

In 2009 he chaired the 10th FIGIPAS Meeting in Inorganic Chemistry held in Palermo on July 2009. Since 2004 he organizes the prestigious Seminar "Marcello Carapezza" whose 2011 edition was hosted by Italy's Parliament.

Co-author of 20 books, Dr Pagliaro is amongst Italy's most cited scientists in nanotechnology and materials science (> 6,400 citations, *h index* = 37, as of early 2017).

In 2015 he co-edited (with Paolo Fornasiero) the "Pd Catalysis" thematic issue of *ChemCatChem*, following the 2014 "Hybrid Materials" *Nanoscale* themed issue (guest-edited with Jean-Marie Nedelec) and the 2011 "Heterogeneous Catalysis for Fine Chemicals" issue of *Catalysis Science & Technology* (along with Graham Hutchings). In 2017, he is co-editing the "Supported Molecular Catalysts" thematic issue of *ChemCatChem* (with Bert Sels).

Reviewer for most major journals in chemistry, energy, and materials science, as scientific expert Dr Pagliaro regularly evaluates proposals for leading research agencies, including the French Research Agency, the Royal Society, the Israel Science Foundation, the Pazi Foundation, the US-Israel Binational Science Foundation, Estonian Research Council, Poland National Science Centre, ACS Petroleum Research Fund, and German Academic Exchange Service.

He is a member of the Advisory Board of *Chemical Society Reviews, Energy Science & Engineering, ChemistryOpen*, and *Sustainable Chemistry and Pharmacy* as well as of the Editorial Board of *Silicon, Coatings* and *Letters in Organic Chemistry*. Along with Vania Zuin, he co-edits the Green and Sustainable Chemistry section of *Chemistry Central Journal*.

In *Flexible Solar Cells* (Wiley-VCH, 2008) he introduced the term "Helionomics." Between September and December 2015 he was president of Palermo's energy utility AMG Energia. His website (www.qualitas1998.net) is online since year 2000.

Preface

With more and more stringent environmental concerns backed up by public awareness, the chemical industry has recently tasted the flavor of its next biomass revolution, which will have anyway to be achieved when the available quantities of fossil resources become insufficient to fulfill the Humanity's demand. However, the political uncertainties (unexpected modifications of regulation policies, etc.) and a barrel price unpredictably jumping up and down, cloud over the visibility of the potential of biosourced products.

The biodiesel industry is not an exception to the rule. While biodiesel production was until recently somewhat lucrative in Europe thanks to favorable incentives, it is nowadays at stake notably due to a turnaround on biofuels policies, which further globally makes the chemical industry dubious, in turn making investors risk averse.

As a catch point, a biodiesel unit produces about 10% of glycerin as a by-product. The earlier biodiesel boom motivated the industrial and academic community to focus research activities on glycerol (refined glycerin) chemical upgrading to value-added applications. To get rid from regulations uncertainty and to subsidies fluctuations, the fate of the biodiesel industry can be profitably interwoven with that of glycerol. In an optimized biorefinery concept, the rationalized production of multiple products is a clear strength, and, among others (seed meal, etc.), coproduction of glycerol-derived high value–added products together with biofuel (here, biodiesel) could provide solutions for a globally equilibrated industry, at least from an economical point of view.

In the aforementioned context, a damper was, however, put on the chemical research associated to the biodiesel industry, in a vicious circle, where the forecasted lower biodiesel production quantities thus generate lower quantities of usable glycerol than expected. Then, research on glycerol, upscaling activities and industrial developments were unfortunately quite put on hold. However, boosting the glycerol upgrading activities can be considered as a good lever to reactivate the whole biodiesel industry, through a rethinking of the associated philosophy. Thus—together with upgrading of fatty compounds to various value-added chemicals—considering glycerol as the main product of a "biodiesel" unit might be one of the keys for a healthier biodiesel industry.

The present book is a gold mine for those who are looking out for a better economical context while proactively preparing an offensive with optimized glycerol-derived lucrative products. The state-of-the-art of the technologies developed so far are described from a scientific and an economical points of view. This piece of work is thus not only precious for deciders, but also for students, and for scientists/engineers to catch some opportunities and to breath inspiration in this field that will undoubtedly be of the upmost importance in the mid-term.

Those who would have been educated, trained in the field, and who would have mastered knowledge and knowhow upfront, will have a considerable advantage when it will be timely to implement glycerol-based advanced technologies. This book is a must-read item for getting comprehensively prepared to the unavoidable boom of glycerol-derived products.

I personally congratulate the author for this initiative, and for the quality of this piece of work, I am myself considering as a landmark in the field.

Franck Dumeignil
Université de Lille 1, Villeneuve-d'Ascq, France
Unité de Catalyse et Chimie du Solide,
UMR CNRS 8181, Villeneuve-d'Ascq, France

Acknowledgments

My deep gratitude goes to Rosaria Ciriminna, Italy's Research Council, for her invaluable cooperation in advancing our joint research in many fields of contemporary chemistry. Thanks to Professor Michele Rossi, now retired from the University of Milan, for what I learned during several years of successful cooperation in the field of glycerin valorization, and for his genuine friendship.

I am indebted to Leela Landress, ICIS, for helpful discussion on many of the topics on this book, and Doris de Guzman, Tecnon OrbiChem, for sharing her vision on the future of glycerol in the context of the bioeconomy. Doris' work has been a constant source of inspiration also through her tireless blogging activity on bio-based chemicals (Green Chemicals Blog).

Numerous scientists in academy and industry shared their expertise and plenty of valued information, including pictures, helping to produce a practically useful work. In particular I wish to thank Gadi Rothenberg (University of Amsterdam), Jeffery Mahaffey (Glycos), Franck Dumeignil, Sébastien Paul and Benjamin Katryniok (Ecole Centrale de Lille), Cristina Della Pina (University of Milan), Sebastián Vásquez Bonilla (University of Panama), Laura M. Ilharco and Ana C. Marques (University of Lisboa), Alexandra Fidalgo (Universidade Europeia), François Jerome (Université de Poitiers), Thibaud Caulier (Solvay), Paolo Bondioli (INNOVHUB), and Serge Kaliaguine (Université Laval) who invited me to give a lecture on glycerol chemistry at the 22nd Canadian Symposium on Catalysis (Quebec, 2012).

Finally I would like to thank Katey Birtcher, Senior Acquisitions Editor, Emily Joy Grace Thomson, Editorial Project Manager, and Anitha Sivaraj, Elsevier, for their excellent assistance in devising and then producing this book.

Mario Pagliaro
Palermo, November 2016

Introduction

Consider the following: Versalis, the leading petrochemical company in Italy, in 2015 had revenues for €4.83 billion [1]. The company has about 4000 employees. These two figures translate into a generation of value of almost €1.21 million per employee.

Two reasons explain such huge value creation. One is that petrochemical industry uses highly efficient heterogeneously catalyzed processes to produce chemicals and polymers. The other is the low cost of oil, the industry's raw material, which lasts since early 2014.

This is the industry with which the emerging bioeconomy companies wishing to produce bio-based products will compete. Perhaps this explains why in the course of the last 2 years numerous companies targeting bioproducts such as bioplastics were forced to bankruptcy, were acquired by petroleum companies or were forced to change business models targeting higher value products such as cosmetic, pharmaceutical, or nutraceutical ingredients.

Glycerin, currently produced in 4 million tonnes per year amount, almost entirely stems as by-product of biodiesel and oleochemicals (including soap). Both industries are greatly affected by low oil price. Indeed for the first time in 12 years, in 2015 the global glycerin output slightly declined.

However, we argue, biomass-based chemical productions, including those relying on this highly versatile molecule, are far from being doomed.

The combination of low and rapidly declining EROI (energy returned on energy invested) for oil, demography, and global economic growth demands some 32 additional million barrels per day by 2025 [2].

In other words, in less than a decade, for the global economy continuing to grow at natural pace, mankind should add more than one-third of current 90 million barrels consumed every day. For comparison the highest amount of additional oil during the US shale oil boom was around 5.5 million barrels per day, and since then the output has declined by 1 million barrels per day (18%) in 18 months.

Put simply switching the production of chemicals and polymers to biomass, and that of energy to renewable energy sources, are the only options left prior to serious global energy and resource crisis.

Current low oil prices are the outcome of global economic depression, which also commands unprecedented low price for refined glycerol and 80% crude glycerol.

About one decade ago epichlorohydrin (ECH) became the first chemical to be manufactured from glycerol as two petrochemical companies in Europe retrofitted existing plants (one in Czech Republic and the other in France) from making glycerol from ECH, to making ECH from vegetable glycerol.

In the subsequent 10 years numerous chemical and biotechnology innovations have been introduced thanks to which glycerol can be conveniently used to manufacture valued chemicals, polymers, and functional formulations.

Following a presentation of the history, market and main applications, the innovative uses of glycerol are presented based on real market success cases (and failures) as well as guided by what we believe will be the forthcoming winning applications.

Easily obtained via Fisher esterification reaction between glycerol and citric acid, for instance, Rothenberg's and Alberts' Plantics-GX is the first thermoset resin capable to penetrate a huge market dominated by the highly conservative plastic industry.

Will it make it?

The answer to this and related questions for other glycerin derivatives requires a closer understanding of the oil, demography, and economy dynamics, which are presented in the last book's chapter.

Using a succinct style this book combines both real market applications as well as prospective for future uses, to convey data and analysis that will hopefully assist the work of researchers and managers working at the introduction of successful bio-based products using glycerol as platform chemical or main ingredient.

Oleochemical and biodiesel manufacturers in what Brunskill has called "the late palmian era" [3] need to proactively act to valorize their main coproduct.

First they need to refine crude glycerin up to pharmaceutical and food grades. This is what happening at several biodiesel and oleochemical plants where new glycerin refineries have started to operate in the last 2 years in the United States, Germany, Malaysia, Singapore, and several other countries.

Then, beyond selling refined glycerin, the very same companies need to start to use glycerin as a platform chemical to efficiently make valued chemicals and formulations in high demand, marketing the new products based on price and performance that most often includes ability to quickly biodegrade and the absence of toxicity. Examples include polyglycerols, biopolyols for the polyurethane industry, antifreeze, and superior cement aid formulations.

In brief, rather than waiting for petrochemical companies to use glycerol as platform chemical or functional ingredient, oleochemical and biodiesel companies should rather start to use their main coproduct to produce a variety of value-added derivatives and formulations using glycerin and glycerin derivatives as main ingredients.

This book will hopefully help these companies and researchers working across the world at the development of significant new uses of "man's most versatile chemical servant" [4].

Mario Pagliaro
Palermo, November 2016

References

[1] Risultati per la prima volta con un positivo utile netto per Versalis, goinpharma.com, February 27, 2016.

[2] F. Meneguzzo, R. Ciriminna, L. Albanese, M. Pagliaro, The Energy-Population Conundrum and Its Possible Solution, arXiv:1610.07298 [physics.soc-ph].

[3] A. Brunskill, An outsider's view of the oleochemical market, POC2015 Palm & Lauric Oils Price Outlook and Exhibition, Bursa Malaysia, March 3, 2015.

[4] M.A. Lesser, Glycerin—Man's most versatile chemical servant, J. Chem. Educ. 26 (1949) 327–328.

CHAPTER 1

Properties, Applications, History, and Market

1.1 AN EMINENT OLEOCHEMICAL

Glycerol (1,2,3-propanetriol) is an eminent oleochemical with unique chemical and physical properties, which originate a wealth of applications. Oleochemicals are chemicals derived from natural oils and fats of both vegetable or animal origin [1].

In the last two decades, both the oleochemical and biodiesel industries have grown steadily due to global demand of greener alternatives to petrochemicals and oil-derived fuels. The oleochemical capacity doubled in 10 years to current 12 million tonnes for acids and over 4.5 million tonnes for alcohols [2].

Glycerol provides the molecular skeleton of all lipids (triglycerides) in which it constitutes on average about 10% by weight of fatty matter from which it is liberated (Scheme 1.1) upon base-catalyzed hydrolysis (manufacture of soap and fatty acids) or transesterification reaction with methanol used to make biodiesel (Fatty Acid Methyl Ester, FAME) fuel.

Generally catalyzed by strong base the transesterification of oil and methanol affording FAME is an equilibrium reaction carried out with methanol in stoichiometric excess (Scheme 1.2).

Since 2004 a large and rapidly increasing surplus of glycerol obtained as a byproduct in the manufacture of biodiesel fuel literally flooded the chemical market. In a concomitant and related trend, the oleochemical industry started to increase its capacity at unprecedented rate.

On June 2012 FELDA Global Ventures Holdings, a Malaysian palm oil and rubber company, raised up \$3.1 billion from the world's second largest initial public offering in that year (after Facebook) [3]. Since then, however, the price of crude palm oil (CPO) has plummeted from >\$1000/tonne to >\$500/tonne in early 2016 urging oleochemical companies to maximize the value extracted from the vegetable oil, including the main coproduct of fatty alcohol and fatty acids production.

Glycerol. DOI: http://dx.doi.org/10.1016/B978-0-12-812205-1.00001-1

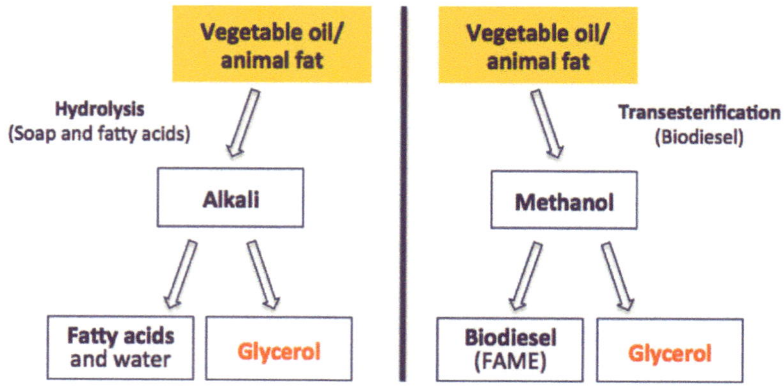

SCHEME 1.1
Both lipid hydrolysis and transesterification reactions yield 10 wt% glycerol as main coproduct.

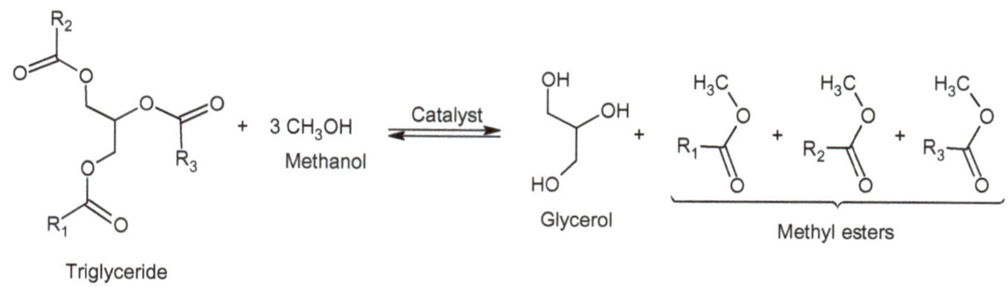

SCHEME 1.2
Transesterification of a triglyceride with methanol.

For comparison the ratio between the 2015 turnover (€600 million) of a leading European oleochemical company and the number of employees (900), [4] translates into €600,000 value generation per employee, namely about half of the value typical of large petrochemical companies mentioned in the Introduction.

Maximizing the value of the products portfolio means to vertically integrate the product value chain downstream.

This is, e.g., what Malaysia's KLK oleochemical producer has done. In 2010 the company bought a plant on the river Rhine in Germany (KLK Emmerich) where it manufactures a range of fatty acids, hydrogenated fatty acids, and glycerol. The initiative was successful and in 2015 the company purchased another plant in Dusseldorf where it produces oleochemicals from both vegetable-based as well as tallow-based lipids, close to key customers and vegetable oil suppliers. In this way the company was only partly affected by low CPO prices, and started to benefit from larger margins due to downstream products of considerable higher value (up to be 50% more lucrative per unit revenue) [5].

1.2 PROPERTIES AND MAIN APPLICATIONS

First isolated and called "the sweet principle of fats" by Scheele in Sweden in 1783 via distillation of the sweet supernatant liquid obtained by heating olive oil with calx lead (PbO) and water [6], glycerol was given its name by Chevreul who read his precise findings at France Académie des Sciences on November 2, 1813 [7]. Another French chemist, Pelouze, derived the empirical formula in 1836 [8].

Glycerol is a colorless, odorless, viscous liquid with a sweet taste, whose name originates from the Greek word for "sweet," *glykys*. In its pure anhydrous condition and under normal atmospheric pressure glycerol has a specific gravity of 1.261 g/cm^3, a melting point of 18.2°C and a boiling point of 290°C (Table 1.1).

The three hydrophilic alcoholic hydroxyl groups are responsible for its complete miscibility with water and its highly hygroscopic nature. This, along with remarkable chemical and physical stability, compatibility with many other chemical materials, nontoxic, nonirritating and environmentally benign nature, explains why the authors of the first 20th century treaty published in 1945, [9] could describe some 1500 different commercial end uses of "glycerin," as the latter term and "glycerine" are widely used in the literature to refer to glycerol.

A highly branched network of molecules connected by hydrogen bonds exists in all phases and at all temperatures. Molecular dynamics simulation suggests that on average 95% of molecules in the liquid are hydrogen bond

Table 1.1 Selected Physicochemical Properties of Glycerol at 20°C

Chemical formula	$C_3H_5(OH)_3$
Molecular mass	92.09382 g/mol
Density	1.261 g/cm^3
Viscosity	1.5 Pa.s
Melting point	18.2°C
Boiling point	290°C
Food energy	4.32 kcal/g
Flash Point	160°C (closed cup)
Surface tension.	64.00 mN/m

Reproduced from CRC Handbook of Chemistry and Physics, 87th edn, Boca Raton (FL), 2006.

connected, [10] leading to unusually high viscosity and boiling point. In detail, glycerol is a highly flexible molecule, forming both intra- and intermolecular hydrogen bonds. Out of 126 possible conformers, calculations indicate that in the lowest energy the hydroxyl groups form a cyclic structure with three internal hydrogen bonds [11]; whereas in the aqueous phase glycerol is stabilized by a combination of intramolecular hydrogen bonds and intermolecular solvation of hydroxyls.

Recently researchers in Iran carried out molecular dynamics to investigate the structural (radial distribution function and pair potential of mean force), dynamical (mean square displacement and transport properties), and thermodynamic properties (density, thermal expansion, and solubility parameter) of glycerol for 293.15–333.15 K at standard pressure [12]. For all temperatures, the simulated densities and viscosities are in a reasonable agreement with the experimental values within 0.4%, the lowest ever reported. The viscosity decreases logarithmically with increasing temperature. The simulated values for viscosity fitted well with the Arrhenius equation ($R^2 = 0.9981$).

As temperature increases, the intermolecular interactions between the glycerol molecules weaken and therefore the viscosity decreases. An Arrhenius-type equation for viscosity, with calculated activation energy of 51.52 kJ/mol in good agreement with the experimental values of 55.76 kJ/mol and 58.33 kJ/mol, shows that the movement of glycerol molecules is a temperature-activated process in which a significant energy barrier that must be overcome before the fluid flows.

As mentioned previously, the unique properties of glycerol are exploited in a broad portfolio of direct uses.

- Pharmaceutical preparations
- Personal care products
- Humectant in tobacco (sprayed on all tobacco leaves prior to shredding)
- Cement multipurpose aid
- Animal feed conditioner
- Solvent (blended with propylene glycol) in liquid used in e-cigarettes
- Glycerin soaps
- Food and nutraceutical/diet low glycemic index ingredient
- Heat transfer fluid and antifreeze.

Refined glycerol is widely used in food, personal, and oral care products where it serves as an emollient, humectant, solvent, and lubricant in an ample variety of products. Glycerol has a very low glycemic index (3; common sugar has 65) and this makes it suitable as part of a diabetic diet as it does not raise blood sugar levels, though it has no particular advantage as a sweetener since it is 0.6 times as sweet as sugarcane and still 4.32 calories/g (in the United States, the Food and Drug Administration requires glycerol to be included in the Total Carbohydrates listed on the Nutrition Facts label).

Remarkably glycerol has unique human metabolism for which there is significant glycerol uptake in muscle tissue but no uptake by adipose tissue [13]. Finally like sorbitol and other polyols glycerol it is not metabolized by bacteria in the mouth that cause plaque and dental cavities. Having become cheaper than sorbitol, glycerol is replacing the former polyol in most toothpastes and mouthwashes.

As ingredient in skin care products, shaving cream, hair care products, and soaps, glycerol acts first as humectant (moisturizer) retaining moisture preventing skin dryness while giving softness.

However, we know today that its biological and biophysical effects go much beyond than simply increasing the hydration of the stratum corneum (SC) [14], to encompass preventive action against the SC phase transition, keratolytic effect by desmosome degradation, increase of the protective function of the skin against irritation and penetration of substances through the SC, plasticizing action toward SC with accelerated healing processes, reduction in tissue scattering, and stabilization of skin collagen.

Refined glycerol is also widely used as a food and beverage additive (labeled E422 in the European Union) acting as solvent, sweetener, moisturizer, and preservative in goods such as baked cereals and dried fruits, which otherwise lose their appeal when they become dry and hard during storage. Hygroscopic glycerol reduces water loss and prolongs shelf life.

Another major direct use is as humectant in tobacco at levels in the range of about 1%–5%. Glycerol is sprayed onto tobacco leaves prior to shredding. The effect of glycerol on the relative deliveries of the various smoke constituents is a substantial increase in water delivery [15]. Vegetable glycerol has lately found another major use, in the smoking industry, as a major component along with propylene glycol of the solvent mixture used to solubilize nicotine in the liquid mixture used in electronic cigarettes. Unfortunately toxic glycidol and acrolein are produced by glycerol dehydration (total aldehyde emissions from 53 to 165 μg/puff by increasing the voltage applied to a single-coil device from 3.3 to 4.8 V) (Fig. 1.1) [16]. Furthermore harmful aldehyde emissions increase by more than 60% after the device is reused several times, likely due to the buildup of polymerization byproducts that degrade upon heating.

The new availability of refined glycerol at low cost and increasing living standards in many countries led to a growth of the use of refined glycerol in personal care from less than 200,000 tonnes in 2002 to 800,000 tonnes

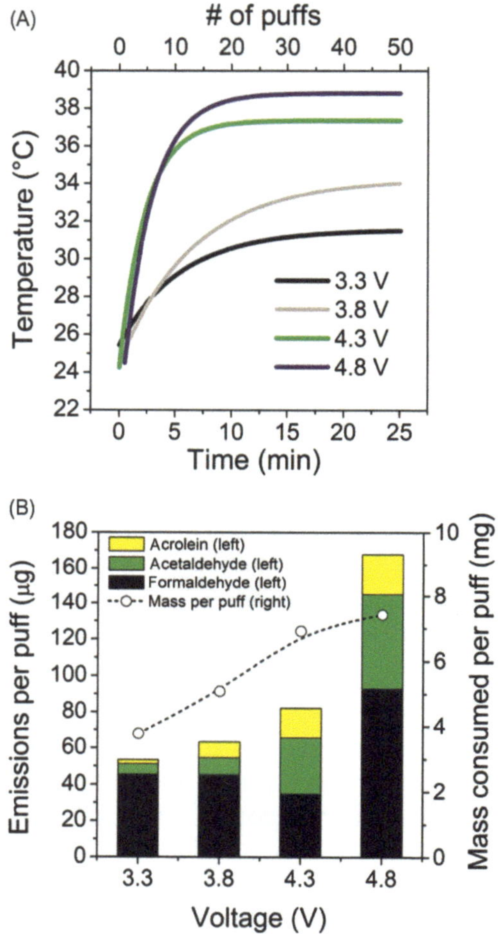

FIGURE 1.1

Effects of the voltage on e-cigarette performance and emission of aldehydes: (A) mean vapor temperature for voltages from 3.3 to 4.8 V (the smoothed curves do not show interpuff fluctuations); (B) aldehyde emissions and mass of e-liquid consumed per puff. *Source: Reproduced from M. Sleiman, J.M. Logue, V. Nahuel Montesinos, M.L. Russell, M.I. Litter, L.A. Gundel, et al., Emissions from electronic cigarettes: key parameters affecting the release of harmful chemicals. Environ. Sci. Technol. 50 (2016) 9644–9651, with kind permission.*

in 2009. Low prices generated new demand from huge countries such as China, Russia, India, and Brazil where glycerol was historically not used due to high price. In 2010 Russia alone absorbed 30% of the total Europe's glycerol export [17]. Similar large figures were recorded in China where consumption went from 100,000 tonnes in 2005 to almost 1 million tonnes in 2014, locally refining crude glycerin imported mainly from Argentina and Indonesia [18].

1.3 AN HISTORICAL OUTLOOK

Glycerol is the raw material for manufacturing the leading explosive nitroglycerin, i.e., the base for dynamite discovered by Swedish industrialist Alfred Nobel [19].

Nobel built bridges and buildings in Stockholm, and while researching new methods for blasting rock he invented in 1863 the detonator for igniting nitroglycerin by means of a strong shock rather than by heat combustion.

Nitroglycerin, the glycerol trinitro glyceryl ester produced by treating glycerol with white fuming nitric acid by Italian chemist Ascanio Sobrero in 1846, in its natural liquid state is a very volatile, heavy, colorless, oily, explosive liquid extremely hazardous to handle. A mere 10 mL will expand 10,000 times into 100 L of gas at an explosive velocity of 7700 m/s—more powerful than TNT. In 1866 Nobel discovered that mixing nitroglycerin with inert absorbent silica kieselguhr (a siliceous sedimentary rock mined in Germany) the hazardous oil discovered by Sobrero could be stabilized turning the liquid into a malleable paste and the resulting explosive handled without danger of immediate explosion.

Remarkably nitroglycerin is also successfully employed since several decades as an antianginal drug, as chest pain suffering workers at Nobel's factories were known to obtain relief when back at work each Monday [20]. Glyceryl trinitrate works by relaxing blood vessels, and the discovery of the biological action of the NO released in the blood upon assumption of NG, leading to a reduction in the frequency and severity of angina pectoris (severe pain in the chest due to inadequate blood supply to the heart) attacks, was awarded the Nobel Prize in Physiology or Medicine in 1998 [21].

Named "dynamite," the new explosive produced little smoke and its secure supply readily became a strategic military issue. Hence when glycerol demand due to war times exceeded the supply from the soap industry, military security reasons led to the first synthetic plants for glycerol manufacture in Europe, Russia, and in the United States.

During World War I synthetic glycerol was produced through microbial, low-yield sugar fermentation. Since 1943, however, synthetic glycerol was obtained from petroleum feedstock thanks to the new high-temperature chlorination of propene to allyl chloride developed at IG Farben plants is Oppau and Heydebreck, in Germany [22]. Since then all synthetic glycerol plants relied on this process.

Until the early 2000s about 25% of the global glycerol demand was met by petrochemical synthesis, and the other fraction from the soap and oleochemical industries wherein glycerol has been a source of additional revenues for about 60 years.

Dow currently operates in Germany (in Stade) 30,000 tonnes/year plant producing 99.7% pure propene-derived synthetic glycerol (tradenamed OPTIM) according to current Good Manufacturing Practices (GMP), aimed at pharmaceutical companies. Yet synthetic glycerol is now a negligible fraction of the glycerol market.

The bulk of glycerol today stems from biodiesel (60%–70% in 2015), and the rest from fatty acids and fatty alcohols production, with a minor fraction obtained from soap making.

Since 2003, indeed, the flood of biodiesel and oleochemical glycerol disrupted a chemical market that had been existing since the early 1940s (Fig. 1.2). Facing low prices and oversupply, most key players in synthetic glycerol pulled out of the market. In Japan the main glycerol production factories ceased operations in October 2005. In the United States Dow Chemical closed its Texas 60,000 tonnes/year plant on January 2006. Two months later Procter & Gamble shut down its 12,500 tonnes/year plant in West Thurrock, London, followed by Solvay which closed its glycerol manufacturing plant in Tavaux, France, and many other producers.

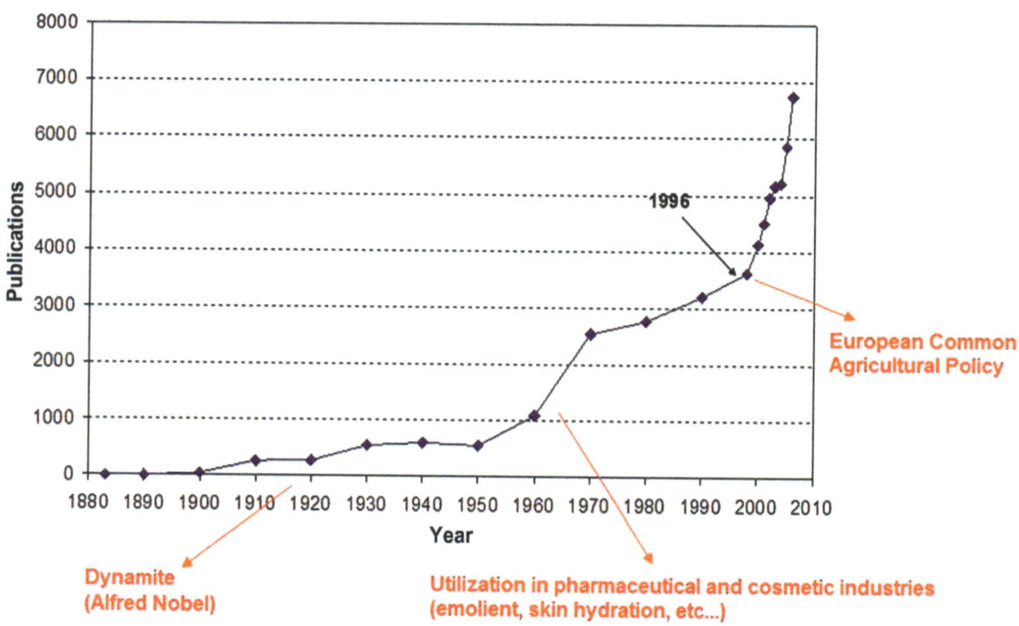

FIGURE 1.2

Trend of publications on glycerol. The number of research papers dealing with new usages for glycerol published between 2000 and 2007 has doubled to more than 7000 annually. *Data from SciFinder database. Image courtesy of Professor François Jerome.*

In May 2003, indeed, the European Directive requiring minimum amounts of biodiesel in diesel fuel (the Biofuels Directive) was published in the EU Official Journal [23]. Subsidies to biodiesel manufacturers were readily enforced as tax exemption in main EU countries with the implementation of mandated minimum volumes of biofuels (5.75% by 2010) in all transport fossil fuels (petrol and diesel). Similar tax exemption and biodiesel blend regulations were adopted in the United States in 2007.

The first biodiesel industrial plant in Europe started operating in Austria in 1985. In 2005 the European Union accounted for 89% of the biodiesel production worldwide [24].

Today regardless of uncertain subsidy policies [25], the United States is by far the world's largest biodiesel producer with about 4.8 million L produced in 2015, followed by Brazil, Germany, France, and Argentine [26].

The technology is simple and capital required to start a biodiesel production plant is relatively low. Hence soon countries with large availability of land such as Argentina, and Brazil or where soy or palm plantation were in place such as the United States, Malaysia, Indonesia, Thailand, and Philippines started to produce and supply biodiesel either for internal consumption blended with diesel (gas oil) fuel, or to export it to Europe. The global production of biodiesel glycerol literally boomed, including a large fraction from increasing oleochemicals manufacturing, even though the quality of raw biodiesel glycerol was very low and, for years, the number of glycerol refineries remained very limited.

Still today (late 2016), Argentina hosts 37 biodiesel plants (10 of which almost closed) processing soybean oil, and only two glycerin refineries: one (Renova) with 110,000 tonnes/year and another (T6 Industrial) of 40,000 tonnes/year capacity [27].

1.4 REFINING BIOGLYCEROL

Today new biodiesel plants operate heterogeneously catalyzed processes that afford pure glycerol as coproduct, such as in the case of a plant which started operation in Nebraska in mid-2016 [28]. Heterogeneous catalysis is much more material and cost efficient than homogeneous catalysis, being the same technology used across the petrochemical industry. In the medium to long term, it will become the mainstream biodiesel production technology. Currently, however, the world's biodiesel industry relies on homogeneous catalysis, and on rapeseed, soybean, and palmitic oils as feedstocks with the most suitable physicochemical characteristics for transformation into biodiesel.

Out of the 31 biodiesel plants built by BDI-BioEnergy, a market leader based in Austria, between 1996 and 2011, 26 of them used KOH as catalyst [29].

Potassium-based catalysts are better than NaOH, particularly for feedstocks with high free fatty acids (FFAs) content (up to 20%), as they ease separation and recovery of soaps and FFAs from glycerol with further yield increase.

Raw vegetable oil is first refined by degumming (elimination of lecithins and phosphorus) and deacidification (elimination of FFAs). The fatty acids are recovered via distillation and sold. Sodium methylate is the best homogeneous catalyst, allowing almost quantitative biodiesel yields [18]; with several plants formerly using hydroxide catalysts having switched to methylate also thanks to lower price and increased availability of CH_3ONa after BASF or Evonik built sodium methylate manufacturing plants in Brazil and Argentina, respectively.

In conventional plants the refined oil is charged into large batch reactors and heated at 55°C with a 30% excess of a mixture of methanol and catalyst. After reaction for 2 h, the mixture is left to stand. The glycerol–methanol solution is heavier than methanol and the esters, and is run off from the bottom of the reactor.

The resulting product, a mixture of biodiesel and water, is dried under vacuum and put into stock pending analytical tests (fundamental parameters are the ester content, minimum 96.5%, and glycerol content, maximum 200 ppm).

Crude glycerol generally contains seven main components beyond glycerol: methanol, water, soap, FAMEs, glycerides, FFAs, and ash. The glycerol content varies depending on the production process. For example, in the analysis of five samples from US biodiesel plants dating back to 2012 the amount of glycerin varied from 22.9% to 63.0% [30].

The crude, furthermore, contains a substantial amount of polyphenols (Fig. 1.3), imparting it distinct antioxidant (and thus anticorrosive) properties, which make it suitable as anticorrosive lubricant and preservation agent [31].

Expensive methanol is stripped from this stream and reused, leaving crude glycerol after neutralization. In its raw state crude glycerol has a high salt and FFA content and a substantial color (yellow to dark brown). Consequently crude glycerin has few direct uses, and its market value, as mentioned above, is often negative.

Perhaps not surprisingly, a number of grease and bioglycerol leakages from biodiesel plant were soon reported, such as in the case of the Alabama Black Warrior river contaminated with 450 times higher than permit levels for two miles downstream in 2008 [32].

To find use in food grade and pharmaceutical applications, crude glycerol must be purified at least to 99.5 wt% purity. Furthermore high grades meeting requirements such as those of the United States Pharmacopeia (USP) or Food Chemicals Codex (FCC) must be produced in refineries that conform

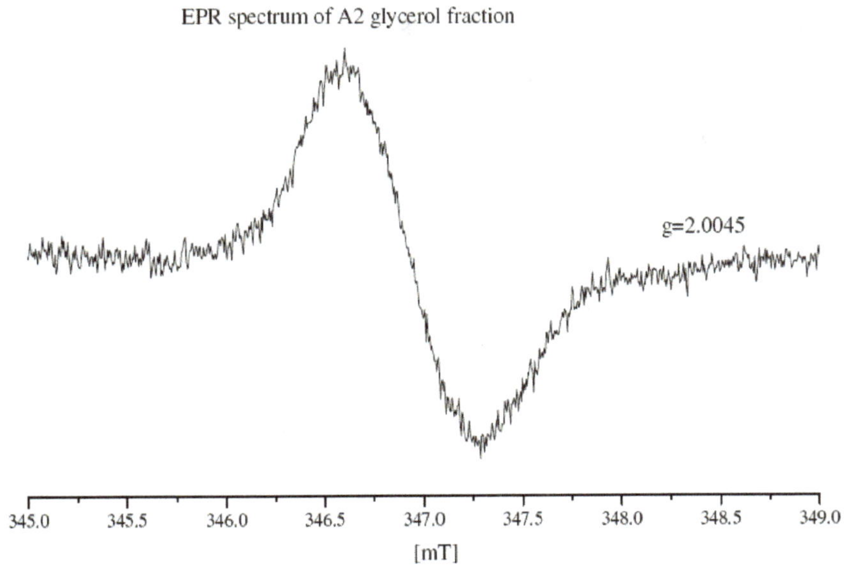

FIGURE 1.3
The EPR spectrum of glycerol fraction shows the typical phenolic/semiquinone radical signal characteristic of an unpaired electron centered on an oxygen atom substituted to an aromatic ring. *Source: Reproduced from M. Jerzykiewicz, I. Cwielag, W. Jerzykiewicz, The antioxidant and anticorrosive properties of crude glycerol fraction from biodiesel production, J. Chem. Technol. Biotechnol. 84 (2009) 1196–1201, with kind permission.*

to strict requirements that, in the case of USP grades, undergo also inspection by the FDA. In general, furthermore, neither the religion of Islam nor Judaism allow consumption of tallow-based glycerol. Hence food grade, vegetable glycerol that has been prepared and maintained in compliance with the customs of the Jewish or Islamic religion often undergoes Kosher or Halal certification.

Refined glycerol is classified into three main classes (Technical grade, USP, Kosher/Halal; Table 1.2).

Traditional purification of raw-glycerol employs high temperature, vacuum distillation. Basically the purification process (Scheme 1.3) starts with neutralization with hydrochloric acid, in which also the cationic component of the transesterification catalyst is incorporated as sodium or potassium chloride.

Following filtration and centrifugation, glycerol is distilled under vacuum at strictly controlled temperature and pressure conditions to avoid decomposition or oxidation, and eventually isolated as food or pharmaceutical grade glycerol following a finishing step over activated carbon to reduce residual color, fatty acids, and other minor components.

Table 1.2 Glycerol Specification Based on Grade and Application

Type of Glycerol	Glycerol Content (%)	Sources/Applications
Crude glycerol	80%	Coproduct of biodiesel or oleochemical production
Technical grade	>96%	Suitable for industrial application, used as a building block in chemicals, not used for food or drug formulation
United States Pharmacopeia (USP)	99.5% USP (tallow-based)	From animal fat or plant oil sources, suitable for food, pharmaceuticals, and cosmetic products
USP and Food Chemical Codex (FCC)	99.7% USP/FCC 99.5% USP (vegetable-based) Kosher/Halal	From vegetable sources only, suitable for use in Kosher/Halal food and beverage

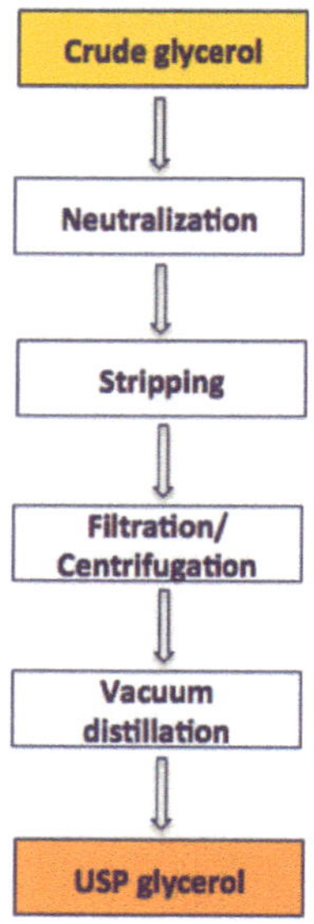

SCHEME 1.3
Steps in crude glycerol purification via vacuum distillation.

Reviewing progress of crude glycerol purification technologies using various techniques in late 2014 [33], researchers in Malaysia concluded that for the next few years vacuum distillation, albeit costly and energy intensive, will remain the main purification method; but that alternative purification methods such as membrane separation technology will eventually emerge once problems such as the fouling and the durability of membranes will be solved.

1.5 A MARKET IN DISARRAY

The unique feature of the glycerol market is that since more than a decade, the bulk of glycerol supply is entirely independent of market demand, providing one of the few examples of a good whose price is *not* affected by the demand for various end-use segments, because there is as much glycerol as much vegetable and animal fats are actually converted into biodiesel and oleochemicals.

Indeed while in 2005 the top three global glycerol suppliers were Procter & Gamble, Cognis, and Uniqema (now Croda) that combined had more than one-third of the market share, 5 years later the main glycerol suppliers were biodiesel and oleochemical companies mostly based in Southeast Asia (Malaysia, Philippines, Thailand, and Indonesia). Today the market has further consolidated with four major companies (IOI group, Wilmar International, KL Kepong, and Emery Oleochemicals) accounting for >65% of overall ($2.47 billion) market in 2015 [34].

Volatility, that once was linked to weather and to fluctuating *demand* of soap and fatty alcohols mainly used by personal care and pharmaceutical industries, now originates from the volatile nature of the glycerol *supply* which, in its turn, is influenced by two main factors: politics (i.e., fiscal incentives to biodiesel manufacturing, biodiesel blend mandates, and subventions to oleochemicals manufacturing and exporting companies), and the price of crude oil.

The only two products to have a value, and thus a price, are 80% crude glycerol having undergone a first purification step after biodiesel and oleochemicals production, and highly pure >99.5% glycerol. Showing evidence of the sensitive dependence of price on the biodiesel global output, Fig. 1.4 shows that the dramatic fall in price of 80% crude culminated in $50/tonne as of mid-2006 was followed by a surge in demand due to ultralow prices.

This in its turn caused a quick price increase reflected in the $830/tonne price of March 2008 following temporary lower biodiesel output in Europe, and poor quality of crude glycerol imported from the United States where biodiesel production was booming and glycerol refineries were still lacking [36].

In the subsequent 8 years, the bioglycerol output continued to grow at much faster pace than new application absorbing the glycerol surplus, and the price plummeted again.

Despite frequent warnings from market analysts emphasizing reliance of biodiesel on tax credits, the biodiesel production never ceased. The overall glycerol output in 2015 amounted to 4 million tonnes, with a significant fraction originating from the oleochemical industry, mainly from fatty acids, followed by fatty alcohols and soap making. In late 2016 spot prices of crude glycerol cargoes going to China were assessed at $190–200/tonne [37].

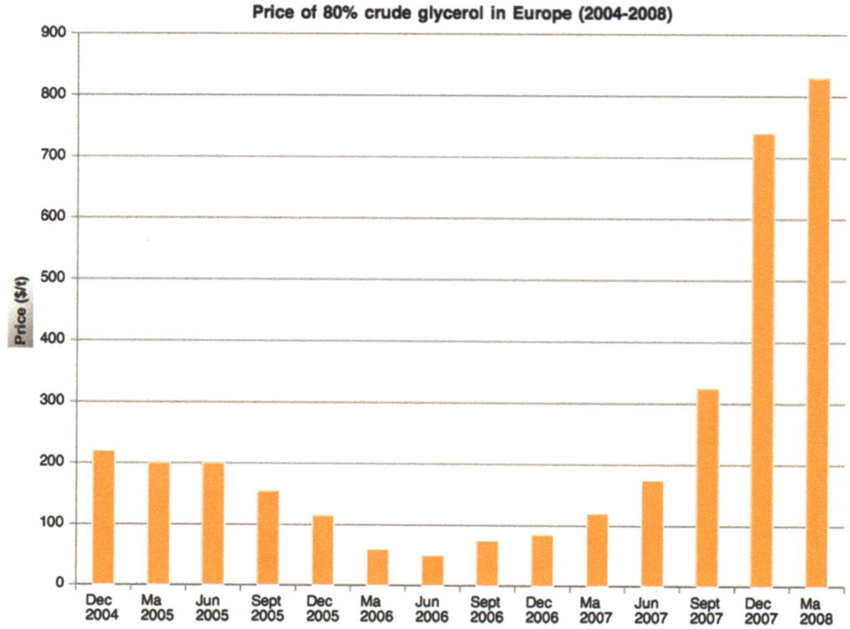

FIGURE 1.4

Price of 80% crude glycerol in Europe (2004–2008). *Source: Produced with data from Ref. R. Ciriminna, M. Pagliaro, in: D. Atwood (Ed.), Sustainable Production of Glycerol in Sustainable Inorganic Chemistry, John Wiley & Sons, New York, 2016, pp. 119–126.*

1.6 PYRAMID OF VALUE

In 2015 an international conference was organized in Paris by a leading glycerol brokerage company, which in 2015 intermediated 400,000 tonnes of glycerol (HB International), to discuss how to solve the glycerol oversupply issue via better balance between supply and demand. The conference brochure read [38]:

> As you know, the glycerine market is in a state of disarray at the moment. Refined glycerine prices in China and Europe (by far the largest markets globally) have approached record lows. Many thousands of tons of crude glycerine have been disposed of at negative values in 2014.

As mentioned previously main glycerol direct uses concern both its use as functional additive in the food, tobacco, and pharmaceutical industries. Developing new chemical uses of glycerol as platform chemical is needed to lower the overall biodiesel production cost [39] finding remunerable outlets for a substance, crude glycerol, that otherwise will find low value-added utilizations as road antidust, deicing [40] agent, cattle [41] or bacterial [42] feed, and concrete additive [43].

In the 2008 book coauthored with Michele Rossi we were writing that "in three to five years glycerol will be seen as an environmentally friendly way of replacing competing petroleum products" [44]. The forecast turned out to be true and today many of the glycerol derivatives mentioned therein, including epichlorohydrin (ECH), propylene glycol, and 1,3-propanediol, are commercially manufactured in large amount.

The development of economically viable new catalytic routes to valued glycerol derivatives has paid off. Valued products such as ECH and mono propylene glycol (MPG) manufactured on industrial scale in southeast Asia, Europe, and the United States have been for years the most successful examples of profitable bio-based products [45].

In general it is the position of glycerol-derived products in the pyramid of value in Fig. 1.5 to indicate whether the product or specific direct utilization will continue to generate profits with glycerol price fluctuations.

Products such as ECH, glyceric acid, dihydroxyacetone, and similar derivatives obtained by selective catalytic or biocatalytic processes sit on top of the pyramid. Their value is significantly higher than that of crude or refined

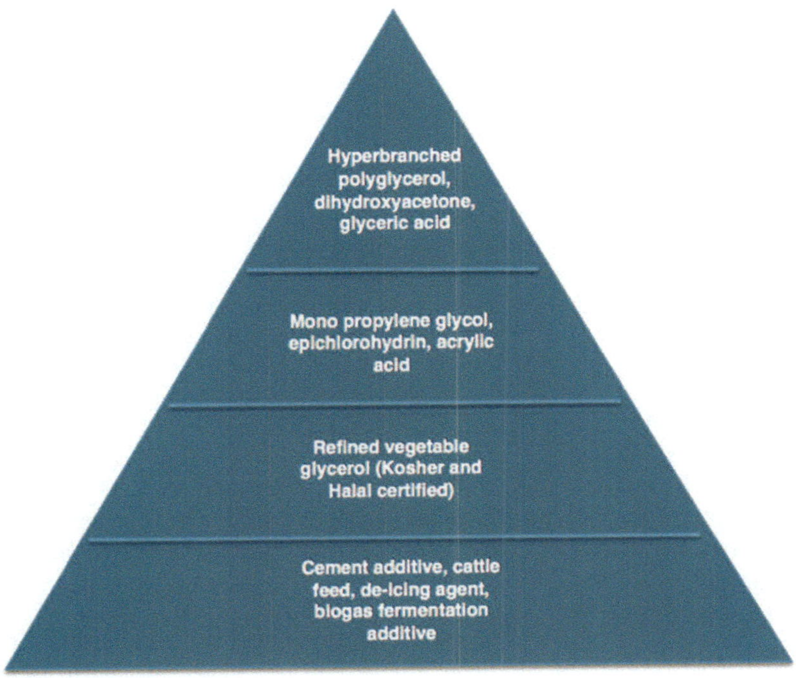

FIGURE 1.5
The pyramid of value for glycerol derivatives and usages.

glycerol, and the impact of price increase will be reduced. On the other hand all products sitting at the bottom of the pyramid—e.g., crude biodiesel glycerol as cattle feed—are greatly affected by price increases, as customers will readily shift to the less pricey alternatives when the price of crude glycerol goes above a certain threshold.

There are many other examples of valued chemicals that can be conveniently obtained from bioglycerol. Asked to comment, longtime analyst of the green chemicals market Doris de Guzman, adds [46]:

> Despite the current global oversupply for glycerine due to increasing biodiesel production, demand for this chemical building block remains robust due to its varied applications. The industry continues to look for new uses for glycerine especially for crude biodiesel-based grades, and there has been success for new applications such as production of epichlorohydrin and propylene glycol.
>
> However the current low petroleum oil situation has slowed down some of these research especially in the development of glycerine-based acrylic acid, methanol and even 1,3-propanediol.
>
> R&D for new glycerine usage continues and there are hopes for applications in higher value albeit lower volume specialty markets. The glycerine industry is also monitoring the rise of renewable diesel production, which is now competing against fatty acid methyl ester (FAME) biodiesel and unlike the traditional biodiesel production, renewable diesel does not produce the co-product glycerine.

Hence the first question for any company considering to switch to glycerol as platform chemical is to understand whether glycerol-based routes will be still competitive with propylene-based processes in the near to medium term.

Will glycerol oversupply continue? And, in that case, at which extent?

1.7 A LOOKING AHEAD PERSPECTIVE

To answer the previous two questions it is instructive to look back at the economic and energy context in which competition between glycerol and oil-derived platform chemicals such as propylene will take place. The future remains uncertain and every forecast is intrinsically likely to be wrong. Nonetheless an insight gained from a critical historical perspective has much to teach.

Critics of first-generation biofuels obtained from edible oils, corn, or sugar beet argue that biofuel production is in direct land use competition with food production [47]. The net energy gain of biofuels, furthermore, is very low.

The results of a 2006 life cycle analysis of soybean biodiesel shows that biodiesel yields only 93% more energy than the energy required for its production [48].

Finally the limited environmental benefits of first-generation biofuels, including biodiesel, are challenged by deforestation taking place in Southeast Asia, Colombia, and several other countries to replace tropical forest with palm plantation.

With CPO quoted at about $550/tonne in early 2016, and gas oil selling at around $283/tonne, biodiesel is entirely relying on public incentives decided by governments. The main rationale in subsidizing the biodiesel industry was, and still is, to support the farmer, reduce unemployment, boost economic growth and, in democracies, to win the votes of the beneficiaries.

Through fiscal and direct incentives to biodiesel and oleochemical production, as well as through minimum biofuel blend mandate, governments ensure a growing flow of revenues to the farming companies growing palm (in Asia), soy (in the United States) rape, and sunflowers (in Europe), and to the related oleochemicals and biodiesel industries.

Those governments will likely continue to enforce incentives especially in times of global economic weakness such as those in which the world entered in 2008 with the global financial crisis. Indonesia, e.g., recently raised the mandate biodiesel blend in diesel fuel from 10% to 20% by 2016, with similar policies aimed to boost prices enforced in Malaysia, Colombia, Argentina, and Brazil that today are among the largest FAME manufacturers.

Indeed despite a global recession that has caused a dramatic reduction of the world's industrial output since the global financial crisis started in 2008, the production of vegetable oils to be converted into biodiesel and oleochemicals has continued to grow.

For example, regardless of claimed reduction in biodiesel output [49], the biodiesel production actually *increased* by 10.3% in 2014 [50], with a first (slight) decrease in more than decade recorded in 2015.

Similar arguments are valid for oleochemicals during what Brunskill has called the "late palmian era" in which the oleochemicals industry is "looking a lot like petrochemicals" with "the advent of the oleochemical 'Mega Plant,' costing $500 million and more, and product output measured in the millions of tonnes" [51].

The shift to greener products and ingredients is "inexorable" [52]; with bio-products competing on a cost *and* sustainability basis as "bloggers, brand owners, and big retailers" have become, to say it with Velson, the "new market regulators" [52].

The fall of oil price since 2014, and oversupply of propylene due to new capacity on stream and new technology (from propane dehydrogenation rather than from naphtha cracking) have led to 50% reduction in the price of propylene (from $1300/tonne in early 2014 to <$700/tonne as of October 2015) [53].

China, by far the largest consumer, has even become self-sufficient through the expansion of propylene production as shown by imports in the last part of 2015 altogether eliminated, in comparison to around 750,000 tonnes imported the previous year. Furthermore significant overcapacity is highly likely due to many capacity additions in China, Korea, and India.

In the same 2 years period in which the price of propylene halved, the price of crude glycerol reached unprecedented low levels. Yet concomitant price deflation of both MPG and ECH made the bio-based route uncompetitive with the propylene-based route. The price of oil-derived MPG and ECH, indeed, is the same of glycerol-based MPG and ECH, with sustainability being a factor driving demand only for USP grade propylene glycol.

Very quickly the bio-MPG market share went from 8% in 2013 to 4.5% in 2015 [54]. In Europe, e.g., Oleon stopped to make bio-MPG in its retrofitted plant in Belgium, to rather refine crude glycerol and sell pure grades.

A similar reduction occurred with bio-ECH, with most glycerol-based plants having run at low production rates in the 2014–16 time interval.

In general the future of any bio-based product depends on: (1) feedstock cost and availability; (2) efficiency of the chemical or biotechnological conversion process used to upgrade it, and (3) value of the bio-based product.

Hence the other route to improve the long-term economic sustainability of bio-based routes based on glycerol is to focus research efforts to develop waste-free, low energy chemoselective routes affording products of high value, with a potential large market such as glycerol carbonate, advanced polymers for medical applications, biodegradable plastics, and surfactants.

References

[1] J. Salimon, N. Salih, E. Yousif, Industrial development and applications of plant oils and their biobased oleochemicals, Arabian J. Chem 5 (2012) 135–145.

[2] A. Brunskill, An outsider's view of the oleochemical market, Bursa Malaysia, 3 March 2015. Available at the URL: <http://www.pocmalaysia.com/shared/files/P1_3%20Alan%20B.pdf>, 2015 (last time accessed 22.10.2016).

[3] Reuters, Malaysia's Felda surges 20 percent in debut of world's No.2 IPO, 28 June 2012.

[4] Oleon, Oleon in short: <http://www.in-cosmetics.com/__novadocuments/251146?v=636045285195070000>.

[5] I. Jala, KLK is example of growing from small to giant size company, thestar.com.my, May 16, 2016.

[6] C.W. Scheele, Rön Beträffande et Särskilt Socker-ämne uti Exprimerade Oljor och Fettmor, Kong Vetenskaps Acad Nya Handlingar 4 (1783) 324–329.

[7] M.E. Chevreul, Examen chimique du savon des graisses de porc et de potasse, Ann. Chim 94 (1815) 80–107.

[8] J. Pelouze, Ueber das Glycerin, Journ. f. Prakt. Chemie 10 (1837) 287–293.

[9] G. Leffingwell, M. Lesser, Glycerin: Its Industrial and Commercial Applications, Chemical Publishing, New York, 1945.

[10] R. Chelli, P. Procacci, G. Cardini, S. Califano, Glycerol condensed phases Part II. A molecular dynamics study of the conformational structure and hydrogen bonding, Phys. Chem. Chem. Phys. 1 (1999) 879–885.

[11] C.S. Callam, S.J. Singer, T.L. Lowary, C.M. Hadad, Computational analysis of the potential energy surfaces of glycerol in the gas and aqueous phases: effects of level of theory, basis set, and solvation on strongly intramolecularly hydrogen-bonded systems, J. Am. Chem. Soc. 123 (2001) 11743–11754.

[12] M.B. Moghaddam, E.K. Goharshadi, F. Moosavi, Glycerol revisited molecular dynamic simulations of structural, dynamical, and thermodynamic properties, J. Iran Chem. Soc. 14 (2017) 1–7.

[13] S.W. Coppack, M. Persson, R.L. Judd, J.M. Miles, Glycerol and nonesterified fatty acid metabolism in human muscle and adipose tissue in vivo, Am. J. Physiol. 276 (1999) E233–E240.

[14] J.W. Fluhr, A. Bornkessel, E. Berardesca, Glycerol - just a moisturizer? Biological and biophysical effects, in: M. Lodén, H.I. Maibach (Eds.), Dry Skin and Moisturizers: Chemistry and Function, CRC Press, Boca Raton (FL), 2005.

[15] E.L. Carmines, C.L. Gaworski, Toxicological evaluation of glycerin as a cigarette ingredient, Food Chem. Toxicol. 43 (2005) 1521–1539.

[16] M. Sleiman, J.M. Logue, V. Nahuel Montesinos, M.L. Russell, M.I. Litter, L.A. Gundel, et al., Emissions from electronic cigarettes: key parameters affecting the release of harmful chemicals, Environ. Sci. Technol. 50 (2016) 9644–9651.

[17] The Market Publishers, Glycerol other than crude (including synthetic): European Union Market Outlook 2011 and Forecast till 2017, Birminghan, UK, 2012.

[18] J. Himont, Glycerine market explodes, Green Chemical, January 2014, 6–9.

[19] S.R. Bown, A Most Damnable Invention: Dynamite, Nitrates, and the Making of the Modern World, Thomas Dunne Books, New York, 2005.

[20] J.C.B. Ferreira, D. Mochly-Rosen, Nitroglycerin use in myocardial infarction patients: risks and benefits, Circ. J. 76 (2012) 15–21.

[21] L.J. Ignarro, Nitric oxide: a unique endogenous signaling molecule in vascular biology, Nobel Lecture, nobelprize.org, December 8, 1998.

[22] For a thorough account, see A. Martin, M. Richter, Oligomerization of glycerol—a critical review, Eur. J. Lipid Sci. Technol 113 (2011) 100–117.

[23] Directive 2003/30/EC of the European Parliament and of the Council of 8 May 2003 on the promotion of the use of biofuels or other renewable fuels for transport, Official Journal of the European Union, L123/42-46, May 17, 2003.

[24] A. Demirbas, Biodiesel, Springer, Berlin, 2008, pp. 111–119.

[25] J. Taylor, L. Landress, Outlook '15: Less volatile 2015 expected for oleochemical markets, ICIS News, December 23, 2014.

[26] Statista, The World's Biggest Biodiesel Producers in 2015, by Country (in billion liters), Statista, Inc., New York, 2016, <www.statista.com/statistics/271472/biodiesel-production-in-selected-countries/> (Last time accessed 22.10.16).

[27] M. Spagnolo, Argentinian glycerine market, The 2nd ICIS Pan American Oleochemicals Conference, Miami, October 11–12, 2016.

[28] Duonix Beatrice begins commercial-scale biodiesel production, biodieselmagazine.com, September 13, 2016.

[29] B. Sims, Catalyzing biodiesel growth, Biodiesel Mag. (February 28, 2012).

[30] S. Hu, X. Luo, C. Wan, Y. Li, Characterization of crude glycerol from biodiesel plants, J. Agric. Food Chem. 60 (2012) 5915–5921.

[31] M. Jerzykiewicz, I. Cwielag, W. Jerzykiewicz, The antioxidant and anticorrosive properties of crude glycerol fraction from biodiesel production, J. Chem. Technol. Biotechnol. 84 (2009) 1196–1201.

[32] B. Goodman, Pollution is called a byproduct of a 'Clean' fuel, New York Times, March 11, 2008.

[33] M.S. Ardi, M.K. Aroua, N. Awanis Hashim, Progress, prospect and challenges in glycerol purification process: a review, Renew. Sustain. Ener. Rev 42 (2015) 1164–1173.

[34] Global Market Insights, Global Glycerol Market, Ocean View, Delaware, 2016.

[35] R. Ciriminna, M. Pagliaro, in: D. Atwood (Ed.), Sustainable Production of Glycerol in Sustainable Inorganic Chemistry, John Wiley & Sons, New York, 2016, pp. 119–126.

[36] J. Himont, Marché Global de la Glycerine, 2007-2009, CNRS, Paris: 13 March 2008. See at the URL: <www.cnrs.fr/inc/recherche/programmes/docs/rdr1_13_03_08/heming.pdf>, (last time accessed 22.10.16).

[37] J. Wong, Asia crude glycerine market in stand-off; some sellers keep offers, icis.com, September 08, 2016.

[38] HBI, Global Glycerine Conference in Paris, Paris, May 26–27, 2015.

[39] M.J. Haas, A.J. McAloon, W.C. Yee, T.A. Foglia, A process model to estimate biodiesel production costs, Biores. Technol 97 (2006) 671–678.

[40] R. Sapienza, A.R. Johnson, W. Ricks, Environmentally benign anti-icing or deicing fluids employing triglyceride processing by-products, WO2005030899 A1, (2005).

[41] S.S. Donkin, Glycerol from biodiesel production: the new corn for dairy cattle, R. Bras. Zootec 37 (2008) 280–286.

[42] N. Kolesárová, M. Hutňan, I. Bodík, V. Špalková, Utilization of biodiesel by-products for biogas production, J. Biomed. Biotechnol. 2011 (2011) 16, Article ID 126798.

[43] M. Rossi, C. Della Pina, M. Pagliaro, R. Ciriminna, P. Forni, Greening the construction industry: enhancing the performance of cements by adding bioglycerol, ChemSusChem 1 (2008) 809–812.

[44] M. Pagliaro, M. Rossi, The Future of Glycerol, RSC Publishing, Cambridge, 2008.

[45] G. Bacchini, Market overview and new applications for oleochemicals in Asia, OFI INDIA 2016, Hyderabad, April 14, 2016.

[46] D. de Guzman, Green Chemicals Blog and Technon OrbiChem, Written Communication, October 2016.

[47] R. Rathmann, A. Szklo, R. Schaeffer, Land use competition for production of food and liquid biofuels: an analysis of the arguments in the current debate, Renew. Energ 35 (2010) 14–22.

[48] J. Hill, E. Nelson, D. Tilman, S. Polasky, D. Tiffany, Environmental, economic, and energetic costs and benefits of biodiesel and ethanol biofuels, Proc. Natl. Acad. Sci. USA 103 (2006) 11206–11210.

[49] Agra Renewable Chemicals, 2014 glycerine supply to drop on lower fame production, April 9, 2014.

[50] British Petroleum, BP Statistical Review of World Energy, London, June 2015.

[51] N. Ellard, Overview of oleochemicals 2016 "Tackling Over Capacity", The 2nd ICIS Pan American Oleochemicals Conference, Miami, October 11–12, 2016.

[52] J. Velson, Trends in key end-user markets, The 2nd ICIS Pan American Oleochemicals Conference, Miami, October 11–12, 2016.

[53] Platts, European Petrochemical Outlook 2016, New York, January 2016.

[54] D. de Guzman, New glycerine applications on the horizon, The 2nd ICIS Pan American Oleochemicals Conference, Miami, October 11–12, 2016.

CHAPTER 2

C3-Monomers

2.1 EPICHLOROHYDRIN

Epichlorohydrin (1-chloro-2,3-epoxypropane; ECH) is a clear, colorless liquid with a sweet, pungent odor mostly used to make epoxy resins, also used in water treatment, paper chemicals, synthetic rubbers, and surfactants. Epoxy thermoset resins manufactured by the petrochemical industry are diglycidyl ethers of bisphenol A (4,4′-isopropylidenediphenol; BPA) obtained by condensation of BPA and ECH (Scheme 2.1) [1].

The resins are transformed with a curing agent into cross-linked networks offering excellent corrosion, solvent, moisture, and chemical resistance, high thermal stability (high glass transition temperatures), hardness (high glassy moduli at 25°C), and good adhesion. For these properties, epoxies are widely used to make adhesives, coatings (widely used in the food and electronics industries), and structural parts for the wind turbine, automotive, aerospace, and construction industries.

Global production in 2016 is expected to exceed 2 million tonnes, making up 13.6% of total revenue of the plastics and resins industry ($14 billion in 2016) [2]. Accordingly global ECH demand amounted to 1.53 million tonnes only in 2015 [3]. Huge production capacities of ECH and BPA in China ensure abundant raw material supply [4].

Currently about 85% of the industrial ECH demand is still met via conventional high-temperature reaction of propylene with chlorine followed by hydrolysis. The first step yields a mixture of 30% 1,2-dichloropropanol and 70% 1,3-dichloropropanol (Scheme 2.2).

The high percentage of 1,2-dichloropropanol formed requires further conversion to the 1,3-isomer, which slows down the process. The overall yield eventually is 97% but large amounts of wastewater contaminated with chlorinated by-products such as 1,2- and 1,3-dichloropropane, 1,2,3-trichloropropane, penta- and hexachlorohexane are produced. Indeed in the traditional process for manufacturing ECH from propene, only one of the four

Glycerol. DOI: http://dx.doi.org/10.1016/B978-0-12-812205-1.00002-3

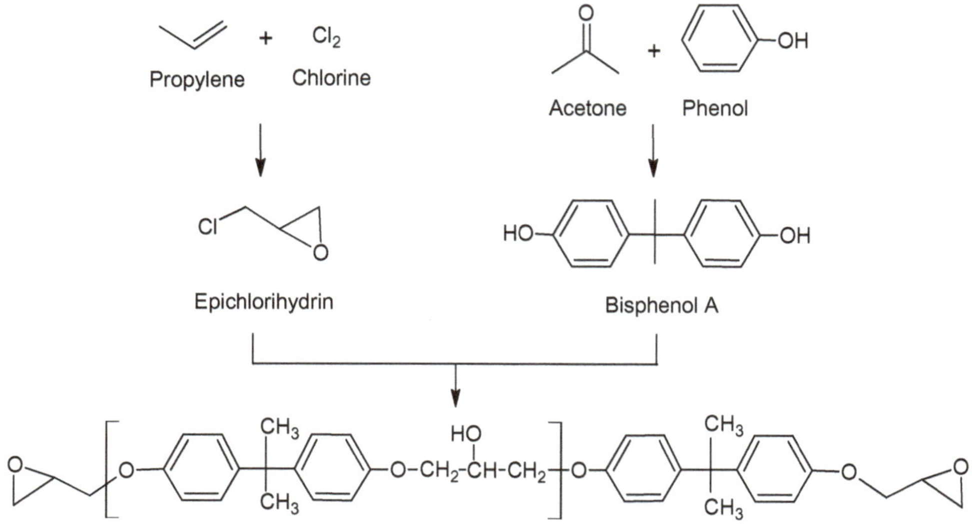

SCHEME 2.1
Manufacturing process for bisphenol A-based epoxy resins.

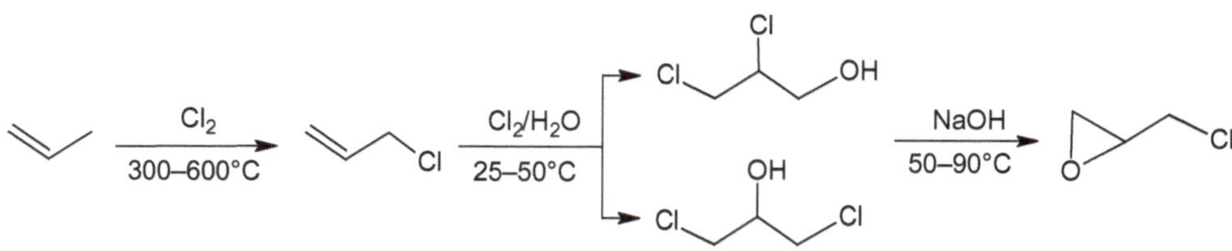

SCHEME 2.2
The traditional industrial process for propene-based epichlorohydrin (ECH).

chlorine atoms involved is retained in the product molecule, the remainder found in HCl and NaCl formed in three of the four reactions involved (Scheme 2.3).

The glycerol-based route reverses the old process (Fig. 2.1), and is used since 2007 to manufacture the remainder 15% of the global output of ECH. Now hydrogen chloride is consumed rather than produced as ECH is obtained via hydrochlorination of glycerol catalyzed by a carboxylic acid, followed by alkaline dehydrochlorination of dichloropropanol.

The three-step reaction mechanism (Scheme 2.4) involves a rate-determining esterification step via nucleophilic substitution on the acylic carbon with

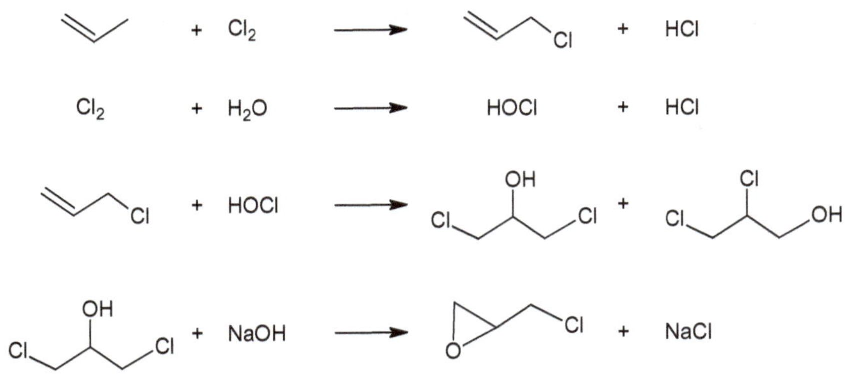

SCHEME 2.3

In the dominant commercial route to (epichlorohydrin) ECH of the four equivalents of chlorine atoms employed, only one is retained in the desired product, the remaining three equivalents appearing as by-product HCl or waste chloride ion.

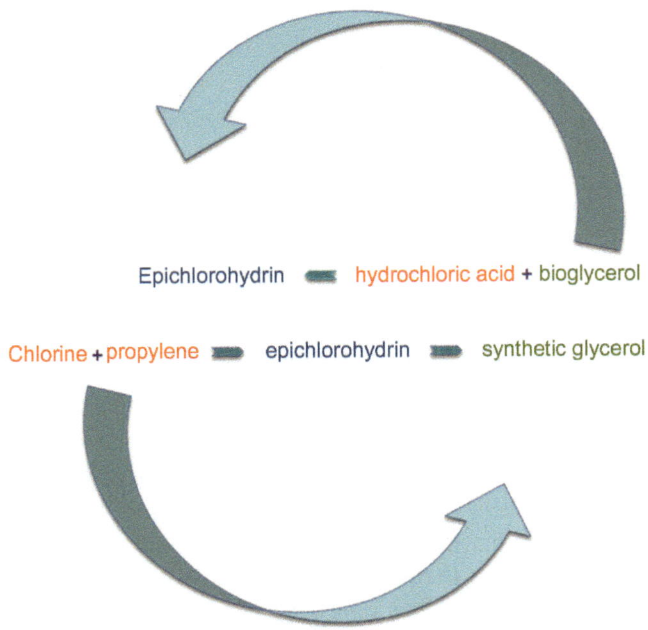

FIGURE 2.1

Hydrochlorination of glycerol reverses the traditional manufacturing process for glycerol from epichlorohydrin (ECH).

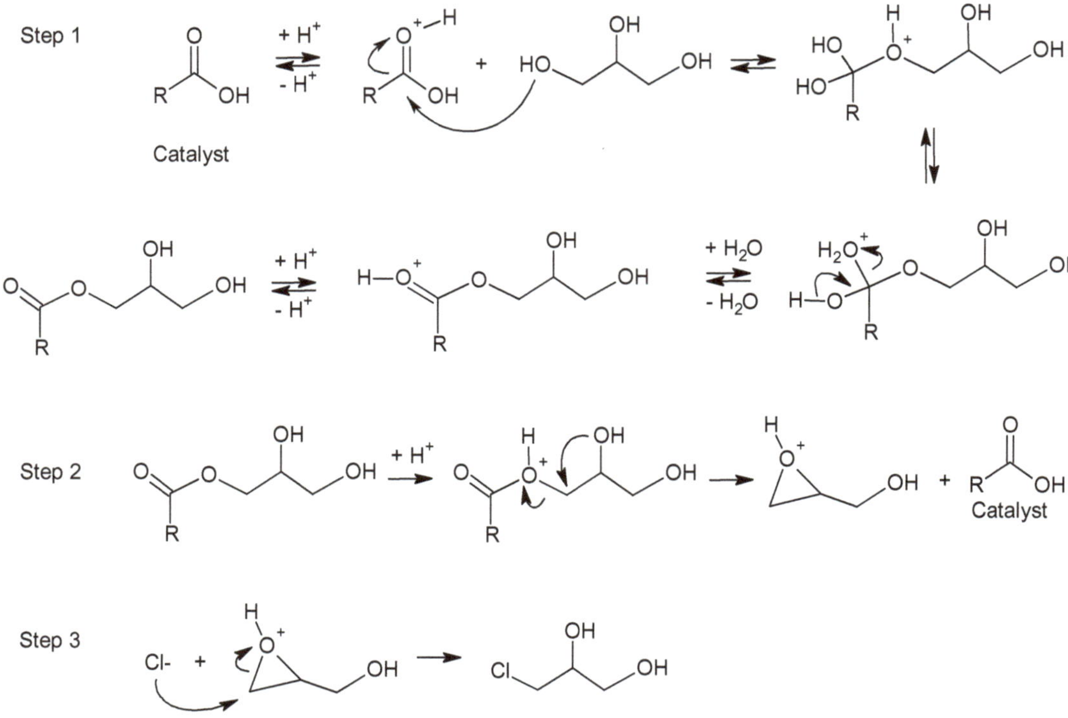

SCHEME 2.4

The three-step reaction mechanism for glycerol chlorohydrin formation via hydrochlorination, explaining the observed reaction path. *Reproduced from reference R. Tesser, E. Santacesaria, M. Di Serio, G. Di Nuzzi, V. Fiandra, Kinetics of glycerol chlorination with hydrochloric acid: a new route to α,γ-dichlorohydrin, Ind. Eng. Chem. Res. 2007, 46, 6456–6465, with permission.*

formation of water, followed by the formation of an oxonium group through alkyl-oxygen bond scission and carboxylic acid release, and the subsequent formation of the chlorohydrin by chloride addition to either the α- or β-carbon atom [5].

The new route comprises initial hydrochlorination of glycerol with hydrogen chloride to give a 30–50:1 mixture of 1,3-dichloropropan-2-ol (α,γ-chlorohydrin) and 2,3-dichloropropan-1-ol, followed by reaction with base to afford crude ECH which is then purified via distillation. Only one equivalent of waste chloride is produced (Scheme 2.5) and practically all glycerol is converted. In the process there is no formation of highly toxic trichloropropanol.

The use of HCl in the gas phase—in the *Epicerol* technology [6] patented by Solvay gaseous HCl is mixed with liquid glycerol and caprylic acid as catalyst at $T > 120°C$—avoids the introduction of water, which has a negative effect on the reaction balance.

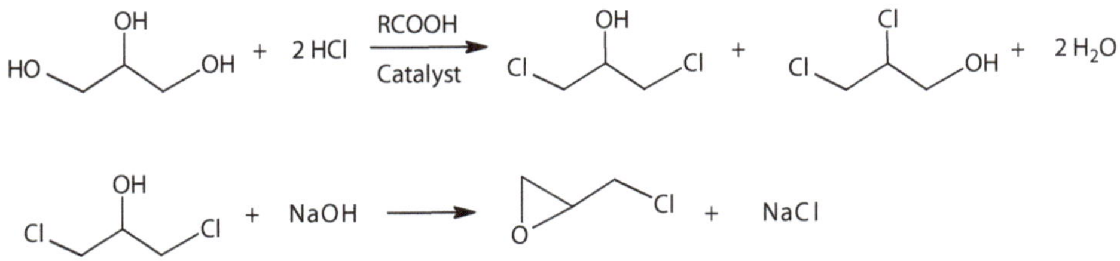

SCHEME 2.5
In the glycerol-based route to ECH only one equivalent of waste chloride is produced.

On the other hand, the use of a catalyst based on long chain carboxylic acids such as eight-carbon caprylic acid, with boiling point above 120°C, compensates for the loss of catalyst due to the reaction temperature, by keeping the concentration of catalyst constant during the reaction.

As mentioned earlier dehydrochlorination using sodium hydroxide generates ECH, whereas the continuous removal of reaction water improves the efficiency and economy of the process [7].

2.2 LIFE CYCLE ANALYSIS OF BIO-BASED EPICHLOROHYDRIN

A glycerol-based ECH plant has a very high intrinsic profitability (about 30%) [8]. Environmentally, too, the glycerol-based route provides several important advantages, though ECH obtained from propene or from glycerol are sold at identical price.

In 1999 Czech epoxy resin producer Spolchemie was among the first companies to reinvestigate the use of glycerol as feedstock to replace propylene in the ECH synthesis. In early 2007 it became the first company to produce ECH from vegetable glycerol using a proprietary process at a new 18,000 tonnes/year plant at Usti nad Labem, in the north of Czech [9].

The same company also produces ECH by the traditional route based on propylene. Comparison between the two routes using the same life cycle analysis standard (ISO) methodology shows that ECH produced from glycerol has generally much lower environmental impact than ECH produced from propylene conventional route: −83% lower figure for global warming potential; −78% lower ozone layer depletion potential; −36% lower acidification potential; and −62% lower photochemical ozone creation.

In 2011 the company received Environmental Product Declaration for its glycerol-based ECH (*Epichlorohydrin G*) by International EPD Consortium [10]. An Environmental Product Declaration is a verified and registered document that communicates comparable information about the life cycle environmental impact of products in agreement with the international standard ISO 14025:2006.

In the Spolchemie process chlorinated organic wastes are incinerated to recover HCl as feedstock for the first reaction step, while wastewater is treated in a bio-treatment plant. In Solvay's *Epicerol* process the aqueous fraction rich in NaCl is recovered and used for the production of chlorine by electrolysis, while water is recycled to provide the aqueous solution needed for hydrodechlorination.

Indeed almost contemporary to Spolchemie, in 2007 Solvay, a large manufacturer of chlorinated products including ECH, retrofitted the plant used until then in Tavaux, France, to manufacture glycerol from propene into a 10,000 tonnes/year ECH plant relying on rapeseed oil biodiesel glycerol supplied from Diester Industries.

Compared to state-of-the-art propylene-based process, the *Epicerol* process shows a 61% reduction in global warming potential due to the carbon capture by the plants (34%) and to the reduction of the greenhouse gas emissions related to the efficiency of the process (not to mention the 80% reduction in formation of chlorinated residues). Moreover, the glycerol-based process does not require a solvent, the size of the reactors is reduced (thanks to higher selectivity), and the kinetics are much faster.

The outcome was so successful that in 2012 the company started the largest bio-based ECH in the world in Ma Ta Phut, Thailand. Operated by Advanced Biochemical Thailand (fully owned by Vinythai, a subsidiary of Solvay), the plant kept producing at more than 70% of its 100,000 tonnes/year nominal capacity.

In 2015 the plant received certification from SCS Global Services for its bio-based ECH from vegetable glycerol [11].

It may not be surprising that the use of renewable glycerin, while beneficial in terms of reduced carbon dioxide and chlorinated waste emissions and energy consumption, contributes to the degradation of other environmental parameters including land use (impact of farmlands), eco-toxicity (impact of plant protection products released in the environment), and water consumption (impact of irrigation) [12]. Solvay's manager Thibaud Caulier was concluding in 2013 that:

> A lot of improvements are still to be expected in the production of glycerin. They are in the crop yield, in the water consumption and in the general farming operations. The bio-based chemical industry can contribute to this evolution, by the development of sustainable sourcing policies involving a close collaboration with raw material suppliers [12].

2.3 PERSPECTIVES FOR BIO-BASED EPICHLOROHYDRIN

Jelle Ernst Oude Lenferink is a process engineering manager at a leading engineering contractor in Europe (Fluor). One of the scenarios lately investigated by his company concerned the possibility to revitalize the European chemical landscape, by reusing existing facilities by making them suitable for using renewable feedstocks, and in particular bio-based glycerol as feedstock to produce ECH [13]. Asked to comment, he added:

> The production of epichlorohydrin using glycerol as feedstock is a known process route, which has been patented and described in several other papers and reports. As always it is about cost, so if you have a cost-advantaged feedstock you are potentially more profitable than the competition, so you make more money etc etc.
>
> Glycerol is a by-product of biodiesel production and locally can have a cost advantage, you prefer not to transport low value components over long distances. Therefore, it is important to know the location of the feedstock compared to the location of the potential epichlorohydrin plant. Building a dedicated, or grass roots plant may not work, however if there is an existing facility that can be modified/revamped, it can be a good option.

Indeed Solvay's subsidiary, Advanced Biochemical Thailand emphasizes its ideally suited location "closer to cheaper feedstock" [14]; similarly to what happens in Tavaux, France, and in Rheinberg, Germany, where Solvay operates another bio-based ECH plants using Europe's biodiesel glycerol (Table 2.1).

Oversupply and slowing global economy forced several glycerol-based ECH producers in China to operate at around 50% or less capacity for the whole 2015 [14], and even before. In addition to the Thai plant, Solvay had plans to invest in another plant of even higher capacity (310,000 tonnes/year) planned to become operational in the second half of 2014 in Taixing, China.

Table 2.1 Main Bio-Based Epichlorohydrin (ECH) Producers

Company	Location	Capacity (tonnes/year)
Solvay	Tavaux (France); Rheinberg (Germany)	50,000[a]
Advanced Biochemical Thailand (Solvay)	Ma Ta Phut (Thailand)	100,000
Spolchemie	Usti nad Labem (Czech)	24,000
Jiangsu Yangnong Chemical Group	Jiangsu (China)	60,000
Yihai Kerry (Wilmar International)	Lianyungang (China) [15]	50,000 + 50,000

[a]*Solvay has two ECH production lines at Tavaux; one that produces ECH from propene and another from glycerol.*

Established in 2012 a new Solvay's subsidiary in China acquired civil construction permit in mid-2013 but shortly afterwards, in the third quarter of the same year, the project was put on hold [16]. Dow Chemical, too, in 2007 announced [17] a forthcoming 150,000 tonnes/year bio-based ECH plant in Shanghai using proprietary glycerol hydrochlorination process technology [18]. To the best of our knowledge, the company has not yet built the plant.

In early 2011 de Guzman, a senior market analyst, was reporting that ever more Chinese ECH makers wanted "to switch to the glycerol-based route to ECH as glycerol cost much less than propylene" [19].

Five years later, with the world's economy facing historic slowdown as shown by plunging Baltic Dry Index, the same analyst was reporting that Asian ECH plants in 2015/2016 were running at 50%–60% average utilization rates of their 659,000 tonnes/year nominal capacity, mostly located in China [20].

Since early 2014 the propylene price has plunged (as of early October 2016, propene was traded at $815/tonne) due to fall in oil prices. So has done the price of glycerol with 80% crude glycerin from Malaysia traded at $240/tonne and refined glycerin at $610/tonne [21].

In brief the ECH market seems to emerge as one eminent example of the inevitable competition between bio-based and petro-based feedstocks for making chemicals and materials.

With two-thirds of the glycerol supply originating from biodiesel, with some 64 countries adopting biodiesel blends including the United States (up to B10 in some states) and Colombia (B10) [22], one might expect that any negative change in the implementation of the biodiesel policies will lead to rapidly decreasing supply of glycerol and high prices.

Conversely, we argue, the Governments of Malaysia, Thailand, Indonesia, Philippines, and other densely populated South East Asian countries will only enhance biodiesel mandates (a B10 biodiesel blend is a mixture of 10% palm methyl ester with 90% petroleum diesel), for self-evident economic reasons.

These policies, indeed, almost immediately translate into higher income for oleochemical and for palm or coconut farming companies in countries with a strong oleochemical economy, and thus ultimately in growth of their national gross domestic product.

Accordingly Indonesia on January 2016 became the world's first country to adopt B20, making it mandatory for vehicles to use fuels with a 20% biodiesel and 80% petroleum diesel blend (planning to implement a B30 policy in 2020) [23]; readily followed by Malaysia which started its B10 biodiesel mandate (from B7) in December 2016. Only in Malaysia, this will translate into a 209,000 tonnes/year increase in palm oil utilization in the 18 biodiesel production plants (from 500,000 to 709,000 tonnes under the new biodiesel mandate) [24].

2.4 TOWARD FULLY RENEWABLE EPOXY RESINS

Many plastics, in particular polycarbonates and epoxy resins, are copolymerized with BPA in its turn obtained via an acid-catalyzed electrophilic aromatic condensation of phenol and acetone using a large excess of phenol to reduce the formation of oligomers.

BPA is a known human endocrine disruptor and its use in polycarbonates and epoxy resins is heavily debated [25], driving the search for a suitable alternative, i.e., both renewable and nontoxic especially for epoxy resins used to protect metal cans from corrosion due to food and beverage such as tomato and beer; with several companies having already removed BPA from their cans.

Two related promising alternatives, enabling to get rid of BPA, have been recently suggested by Caillol's team in France [26], and by Stanzione and coworkers in the United States [27].

Both teams demonstrated the synthesis of epoxy thermosets with good thermomechanical properties from mixtures of phenolics obtained from lignin.

Available in roughly 50 million tonnes extracted each year from the pulp and paper-making industries, lignin is an eminent example of today's wasteful use of bioresources as it is currently burned in about 98% amount, with only around 2% being used not even for high-value derivatives such as polymers and fine chemicals as it would be desirable [28], but rather for low-value products, such as dispersing or binding agents [29].

Vanillin, e.g., is currently obtained in 3000 tonnes/year from spent sulfite liquor, namely by treating an aqueous solution of lignin with sulfites at alkaline pH and high temperatures and pressures, a process that until the 1981 provided 60% of the world's vanillin output, until most lignin-to-vanillin plants closed at the end of the 1990s due to increased environmental concerns toward the caustic effluents of the process, especially the extraction of vanillin obtained in the depolymerization step via bisulfitation with aqueous $NaHSO_3$ [30].

Demonstrating how biomass variability can become an industrial asset for fine product tuning instead of a drawback, Caillol and coworkers used a mixture of phenolics modeling the products obtained from softwood

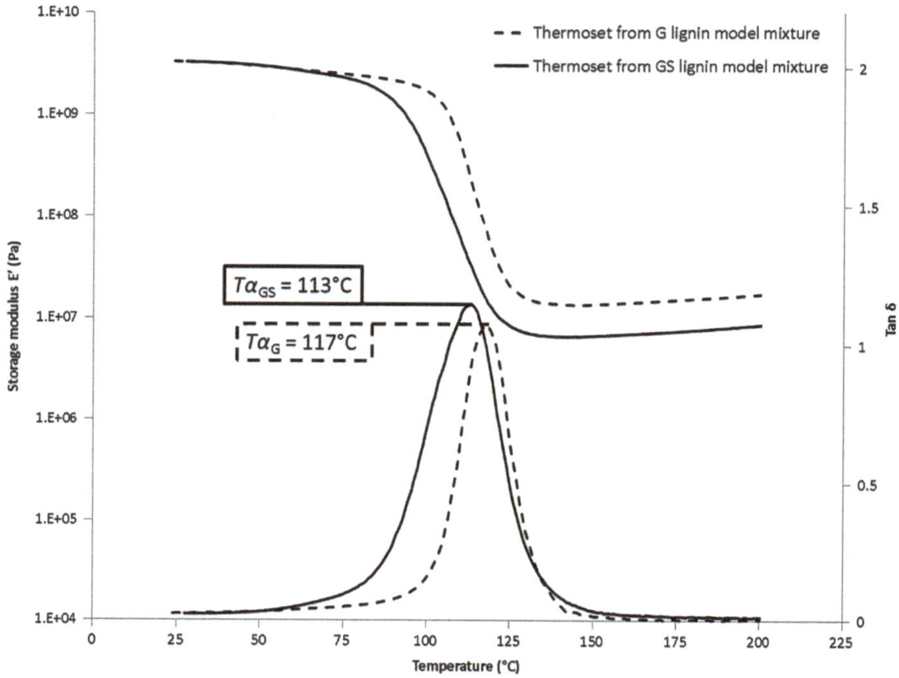

FIGURE 2.2
Dynamic mechanical analyses of the epoxy thermosets prepared from model mixtures of G- and GS-type lignin monomers. *Reproduced from reference M. Fache, B. Boutevin, S. Caillol, Epoxy thermosets from model mixtures of the lignin-to-vanillin process, Green Chem. 18 (2016) 712–725, with kind permission.*

(G, guaiacyl) and hardwood (GS, guaiacyl and syringyl) lignin depolymerization. These mixtures and each of their individual components were first oxidized with sodium percarbonate to increase their phenolic functionality, and then the mixtures glycidylated to obtain BPA-free mixtures of epoxy monomers.

Epoxy thermosets prepared from these epoxy monomer mixtures afforded epoxy resins with excellent thermomechanical properties (Fig. 2.2), with storage moduli $E'(\text{G})$ and $E'(\text{GS})$ at 25°C close to 3.3 GPa, a high-value characteristic of high-performance epoxy thermosets. Remarkably, furthermore, the G monomer afforded thermosets with glass transition temperature slightly higher (117°C) compared to the GS-derived epoxy (113°C), which is consistent with the fact that the mixture of G-type epoxy monomers has a higher mean functionality than its GS-type counterpart, affording a more cross-linked polymer, which is also more homogeneous as shown by the sharper tanδ peak.

The Stanzione's team, in its turn, synthesized a bio-based bisphenolic analogue, bisguaiacol (BG), via electrophilic aromatic condensation of vanillyl

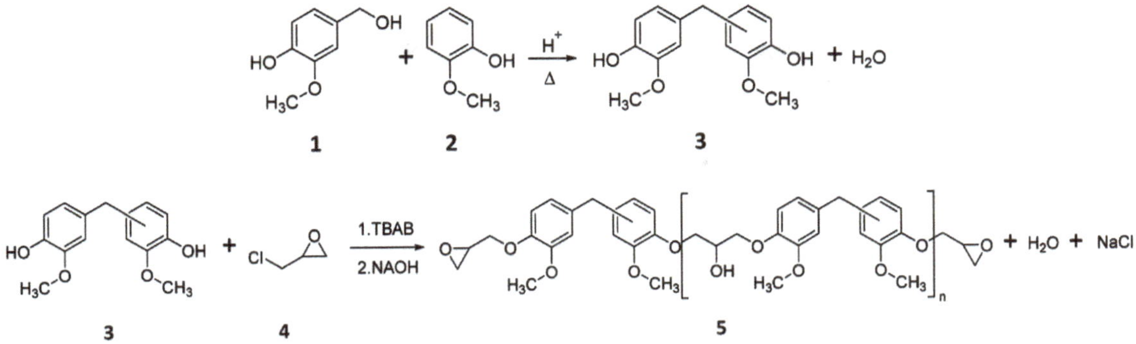

SCHEME 2.6
Electrophilic aromatic condensation of vanillyl alcohol (**1**) with guaiacol (**2**) to produce bisguaiacol (BG) isomers (**3**) (Top); and synthesis of DGEBG (Bottom).

alcohol and guaiacol (two methoxyphenol natural products). The bio-based epoxy BG (3) was then reacted with ECH (4) to produce a diglycidyl ether of bisguaiacol (DGEBG; 5) (Scheme 2.6).

In addition, three single aromatic diglycidyl ethers were synthesized from vanillyl alcohol (DGEVA), gastrodigenin (DGEGD), and hydroquinone (DGEHQ). The epoxy resins were cured, either by themselves or with a commercial BPA-based epoxy resin, with stoichiometric equivalents of Amicure PACM (4,4′-methylenebiscyclohexanamine).

Tested via dynamic mechanical analysis, the bio-based polymers show excellent properties, comparable to the properties of the petroleum-based thermosetting polymers. The thermomechanical results indicate that the methoxy group lowers the glass transition temperature (T_g) yet increases the glassy storage modulus at 25°C, while the methylene spacer between the aromatic ring and the epoxide further lowers the T_g in cured epoxy–amine systems.

Reviewing the most explored bio-sources for bio-based epoxy resins and epoxy resin curing agents (plant oils, lignin, and furanyls whose monomers yield polymers with desirable thermomechanical properties) [31], Stanzione and coworkers lately concluded that lignin is the most important renewable resource capable to provide phenols and phenolic derivatives useful for this significant segment of the polymer industry.

Making products from biomass closes the materials cycle ensuring the basic condition for true environmental sustainability. Bio-derived materials, and polymers in particular [32], become renewable, as it is quickly happening with energy with electricity generation from renewable wind and sun radiation [33].

2.5 PROPYLENE GLYCOL

The selective hydrodeoxygenation of glycerol to C_3-diols 1,2- and 1,3-propanediols is one of the most attractive methods for glycerol upgrading (Scheme 2.7) [34].

Mono propylene glycol (MPG; 1,2-propanediol) is used in the manufacture of high-performance unsaturated polyester resins, as well as engine coolant, antifreeze, deicing agent, and solvent in pharmaceutical and cosmetic products.

The current global production of MPG, 2.18 million tonnes in 2015, was expected to reach 2.56 million tonnes by 2017 growing at a rate of 4.5% per annum [35]. The world market is actually growing at 2%–3% with prices having plunged due to supply from new large production units (e.g., the one in Thailand running since March 2014 with an annual capacity of around 150,000 tonnes/year), significant increase of capacity utilization rates, and global deflation.

For example, by late 2016 MPG was traded at about $975/tonne for the industrial grade and at $1300/tonne for USP grade [36]. As of late 2012 the respective prices for the same grades were $1700/tonne and >$2200/tonne.

We remind that as of late 2016 refined glycerol was traded at about $600/tonne. In general, except for brief periods since 2005, the refined glycerin market price has traded below the MPG price. Hence USP glycerin has actually *replaced* MPG as humectant, solvent and emollient in all personal care products, as well as in all technical applications where possible.

Granted by the FDA the Generally Recognized as Safe (GRAS) status as food and drug ingredient [37], pharmaceutical grade propylene glycol (at least 99.5% pure by weight) has been used safely in many health-sensitive applications such as

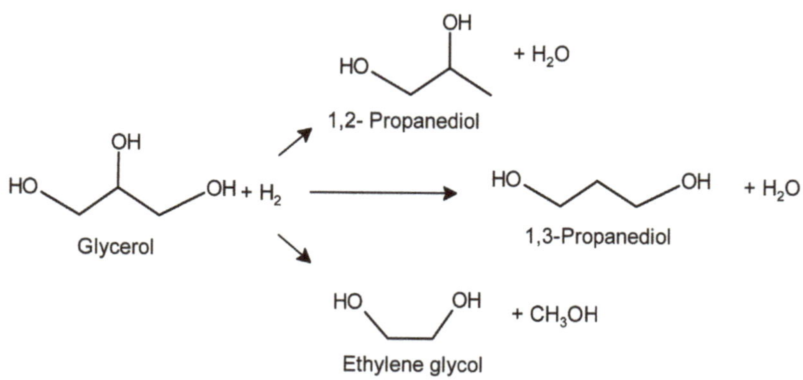

SCHEME 2.7
The conversion of glycerol to propylene and ethylene glycols takes place via catalytic hydrogenation.

food, cosmetics, and pharmaceuticals, even though caution is required when treating patients with large doses or prolonged use of medications containing MPG, especially in infants or patients with renal or hepatic dysfunction [38].

Aptly formulated, MPG is able to effectively lower the freezing point of water to −60°C. Being essentially nontoxic and environmentally benign, thus, during the past decade industrial grade MPG started replacing acutely toxic ethylene glycol [39] as the primary ingredient in the multibillion antifreeze products market.

Alone, antifreeze deicing fluid market is worth >$1 billion [40]. Here replacement of ethylene glycol started from the aviation deicer market (Fig. 2.3), in which MPG is formulated by green formulations free from toxic additives such as triazoles, used as corrosion inhibitors and flame retardants, and nonylphenol ethoxylate surfactants.

MPG is mostly derived from propylene oxide (PO) in a high pressure (20–25 bar), high temperature (120–190°C), noncatalytic hydrolysis of PO using a large excess of water (affording a mixture of 90% MPG and 10% di- and tripropylene glycols).

FIGURE 2.3

Deicing an aircraft at Calgary airport with Type IV fluid deicing fluid. On the average, it takes 3000–4,000 L of fluid to deice a commercial aircraft. *Reproduced from Wikipedia, with kind permission.*

Its production from renewable resources debuted in China thanks to Hong Kong–based Global Bio-Chem Technology that started bio-based MPG production from starch in 2008. Different glycerol-based routes were then commercialized by Archer Daniels Midland (ADM) in the United States, and Oleon in Belgium.

In mid-2011 ADM started MPG production from glycerol at a new 100,000 tonnes/year plant within its large corn wetmill complex in Decatur, Illinois, using catalytic technology developed at Pacific Northwest National Laboratory [41].

In 2012 Oleon transformed a former fatty alcohol plant, and started producing MPG from glycerol at a 20,000 tonnes/year plant in Ertvelde, Belgium, using catalytic technology licensed by BASF [42]. The plant was retrofitted by the engineering and construction division of Air Liquide, which operates a glycerol to MPG pilot plant in Belgium, and is actively marketing the process among oleochemical companies [43].

2.6 HYDROGENOLYSIS TO MONO PROPYLENE GLYCOL

MPG from glycerol is obtained via selective hydrodeoxygenation (hydrogenolysis), namely the removal of an oxygen atom by the addition of hydrogen (Scheme 2.8).

In general Cu-based catalysts are used to mediate the process with almost complete glycerol conversion and 100% selectivity toward MPG [44].

The BASF process used by Oleon, too, uses hydrogenolysis in liquid phase over a copper catalyst using technology developed by BASF in which two serial fixed-bed reactors are used to house the reaction, which occurs at a temperature of 175–195°C under much higher hydrogen pressure (75–200 bar) [43]. The crude product is then sent to a two-column distillation unit to be purified affording pharmaceutical grade MPG.

Another glycerol hydrogenolysis process available for commercialization has been developed by Davy Technologies. The process is a vapor-phase hydrogenation of glycerol, over any conventional hydrogenation catalyst including

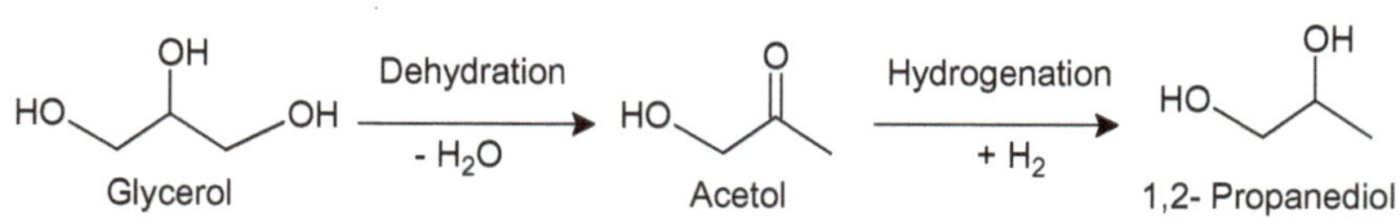

SCHEME 2.8
Hydrogenolysis of glycerol.

heterogeneous copper catalyst, under relatively moderate conditions (20 bar, 205–220°C), affording PG in high yield with remarkably high selectivity [45].

It is surprising that it has been possible to carry out the hydrogenation reaction in the vapor phase since it has generally been believed that this was not possible due to the high boiling point of the glycerol, while the use of high temperatures would be anticipated to cause the glycerol to coke and adversely affect the catalyst.

The glycerol, together with a recycle stream, is vaporized in a recirculating stream of hydrogen. Vapor-phase operation ensures excellent mixing between abundant H_2 and glycerol, achieving ideal catalyst contact. This delivers high conversion rates and temperature uniformity with minimal residence times, which translates into a selective process with minimal by-product formation.

Glycerol conversion is around 99%, and by-products are removed by distillation. In addition, the solid catalyst does not mix with the reactants or products, eliminating any need for catalyst separation prior to downstream processing [46].

Using a large excess of H_2 gas gives it a high partial pressure, which eliminates the need for a high overall operating pressure to drive the hydrogenolysis reaction (as it would happen in liquid phase) and also reduces pipe and vessel wall thickness, reducing equipment, the installation, and other structural costs of the plant.

2.7 THE BUSINESS CASE OF BIO-BASED MONO PROPYLENE GLYCOL

Jeffery Mahaffey is the president of Glycos, a consultancy in oleochemicals, renewable chemicals, surfactants, and chemicals trading. Prior to that, he was chief operating officer of Global Bio-Chem Technology Americas, for which he established a US subsidiary to market bio-based (from starch) glycols and eco-friendly deicing products for the China-based parent company. Asked to comment on the glycerin to MPG route he said:

> Except for brief periods since 2005 when mineral oil prices broke out of their historic price range to the upside, the refined glycerine market price has traded below the MPG price. Most of the direct substitutions of glycerine for MPG have taken place in products such as personal care, detergents, and technical applications where possible. The next logical step is to use the C3 backbone of glycerine to substitute for propylene derivatives such as MPG.
>
> The main issue of converting glycerine to MPG is obtaining a reliable yield. The theoretical yield is 1.22 kg glycerine = 1 kg MPG, though in practice the more typical value would be around 1.3 kg glycerine = 1 kg MPG.

> Secondly, one must secure a reliable source of good quality glycerine at a cost that is below propylene in order to have competitive finished product costs. Crude glycerine has not been proven to work, and poor quality refined will not make an MPG product with sufficient color stability and low odor.
>
> Market acceptance of bio-MPG has been very good. It is for all practical purposes identical to petro-based MPG so buyers do not have to change how they buy. The biobased MPG can be delivered in the same fashion and used in the same system. This gives buyers an ability to both promote higher renewable content in their products, as well as have an alternative during spikes in the propylene market price. If the customer can buy bio-MPG at a price near the level of conventional MPG, he is quite likely to use it.
>
> Bio-MPG represents an interesting new process route, but its bio-nature alone does not create value for the buyer and higher prices relative to petro-MPG. It does create value in that it helps the buyer meet his or her sustainability goals without having to make a lot of changes. It can also create value with consumers who have shown some preference for bio-based products, but this value is captured at the shelf. In a consumer product, for instance, the marketer of that product will want to keep the value he has created by introducing and advertising something that better meets a consumer need and is unlikely to share that with a raw material supplier.

The MPG market is still about 95% petroleum-based. As long as glycerin will remain in oversupply, production of bio-based MPG will continue, even under current low oil prices which translate into low cost of propylene. The typical profit margins of glycerol-based MPG are large and justify bio MPG production as long as refined glycerol prices remain relatively low. Otherwise oleochemical manufacturers will stop making MPG and rather sell refined glycerin, as done by Oleon for a short period between November 2012 and early 2013 because it was then more profitable to sell feedstock refined glycerol, than to convert it to MPG [47].

To capture the extra value of bio-derived MPG, another option is to captively use it as a raw material for added valued derivatives. Again this is what Oleon does supplying the cosmetic industry with "100% natural alternative" propylene glycol esters such as MPG-dicaprylate (Radia 7207) marketed as an excellent emollient having "the same sensorial features as its petrochemical version" [48]. This time, however, the product bears a premium price tag as consumers largely prefer bio-based ingredients.

2.8 BIOLOGICAL ROUTE TO 1,3-PROPANEDIOL

Another glycol of significant industrial interest that can be obtained from glycerol is 1,3-propanediol (PDO). PDO is industrially more valuable than

MPG, but the selective formation of PDO via chemoselective heterogeneous catalysis is difficult as PDO is a kinetically controlled product [34], requiring the switch to biocatalysis.

Reacted with terephthalic acid, indeed, PDO affords the valued polyester fiber polytrimethylene terephthalate (PTT). Described as the "new nylon," [49] actually this fiber offers numerous additional advantages over both nylon and conventional polyester fibers as it possesses a unique combination of chemical resistance, light stability, UV resistance, elastic recovery, softer feel, and dyeability.

Besides the petrochemical route from ethylene oxide (the Shell route) [50], bio-based PDO is produced either from glucose [51], or from glycerol [52].

The glucose-based process, industrialized by a joint venture between DuPont and Tate & Lyle, is based on genetically engineered *Escherichia coli* in which *E. coli* is genetically modified with a gene encoding a nonspecific catalytic activity sufficient to convert 3-hydroxypropionaldehyde (3-HPA) to PDO. DuPont indeed partnered with Genencor to develop the organism that would use the corn-derived glucose to produce PDO.

The proprietary fermentation process, followed by cleaning and distillation, is carried out industrially since 2006 at a 63,500 tonnes/year plant in Tennessee close to a large corn wetmill from which the corn glucose feedstock is pumped to the PDO production facility in large fermenters filled with the microorganism. The broth containing excreted PDO is separated via distillation to form 99.97% pure PDO [53].

The fermentation uses bacterial strains from the groups *Citrobacter, Enterobacter, Ilyobacter, Klebsiella pneumoniae, Lactobacillus, Pelobacter,* and *Clostridium*. The harmless microorganism *Clostridium*, widely disseminated in nature, was shown to convert glycerol to PDO as early as 1881. Since the late 1990s, the microbial conversion of glycerol to PDO has been widely investigated as the process is relatively simple and does not generate toxic by-products.

In general microbes convert glycerol to PDO in a two-step, enzyme-catalyzed reaction sequence. In the first step a dehydratase catalyzes the conversion of glycerol to 3-HPA and water (Eq. 2.1). In the second step 3-HPA is reduced to PDO by a NAD^+-linked oxidoreductase (Eq. 2.2).

$$\text{Glycerol} \rightarrow \text{3HPA} + H_2O \tag{2.1}$$

$$\text{3HPA} + \text{NADH} + H^+ \rightarrow \text{PDO} + NAD^+ \tag{2.2}$$

PDO is not metabolized further and as a result it accumulates in the medium. The overall reaction consumes a reducing equivalent in the form of a cofactor, reduced β-nicotinamide adenine dinucleotide (NADH), which is oxidized to nicotinamide adenine dinucleotide (NAD^+).

In 2005 Liu and coworkers in China reported that simple addition of 5 mM fumarate to cultures of *K. pneumoniae* enhances the rate of glycerol consumption and the production of PDO [54]. In detail using a 5 L bioreactor with 5 mM fumarate addition, the specific rate of glycerol consumption and the productivity of PDO both increased by 35% over the control.

This is the glycerol-based technology that was licensed from Tsinghua University [55], and further industrialized by Zhangjiagang Glory Biomaterial in a 65,000 tonnes/year bio-based PDO production plant in China's Changjiang River Chemical Industrial Park, affording up to 92% PDO extraction rate from glycerol fermentation broth. Almost all of the PDO produced therein is used internally in the production of PTT resin [56].

The bio-route is used by numerous other China's biotechnology companies, including Anhui Lixing Chemical, Fujian Shengda Biotechnology, Chen Bio, Hunan Rivers Bioengineering, and Shenhong Group, which built small-scale biological PDO production lines [57].

Previous research carried out in Germany within the EU-funded research project "Biodiol" had allowed to establish that the use of immobilized rather than freely suspended cells, enabled an increase in productivity from about 2 to 30 g/L h [58] whereas improved process design (fed-batch with pH-controlled substrate dosage) allowed the relatively low product concentration at a maximum of 70–80 g/L (due to product inhibition) to be increased to over 100 g/L.

The European researchers concluded that the classic microbial fermentation of glycerol to PDO was now economically viable, especially because the new bacterial strains isolated can ferment also 80% glycerol, rather than pure glycerol, with the main parameter affecting the bio-based PDO cost being the price of the feedstock employed.

A decade later, state-of-the-art research confirmed that biotechnological fermentation over immobilized cells significantly enhances PDO production enabling continuous production in smaller bioreactors to eventually afford high volumetric product yields in a poorly energy intensive process, while generating little or no waste [59].

In late 2012, METabolic EXplorer, a French biological chemistry company, announced plans to resume construction of a bio-based PDO facility 8000 tonnes/year using crude glycerol as feedstock in the Iskandar region of Malaysia. The facility, already announced in 2010 [60], eventually was not built. The company rather entered into an agreement with Korea-based SK Chemical on July 2014. Engineering studies completed within the partnership confirmed the economic and technical feasibility of the glycerol-based route to PDO [61], even though low ethylene cost likely convinced the Korean petrochemical partner to exit the partnership. Confirming the economic attractiveness, however, a new partnership with oil engineering company Technip was announced in the second half of 2016.

2.9 ACROLEIN AND ACRYLIC ACID

The production of acrolein, another valued C_3 derivative of glycerol, via glycerol double intramolecular dehydration has been the object of intense academic and industrial research efforts in the last decade [62], even though the commercial production of acrolein and acrylic acid from glycerol has not yet been commercialized.

A comparative analysis of petroleum-based and bio-based acrolein production carried out in 2012 concluded that a transition toward bio-based acrolein production was already industrially justified [63], in agreement with a previous process engineering and economic analysis of acrylic acid production from glycerol (Scheme 2.9) compared to the conventional route from propylene monomer [64].

In general dehydration to acrolein is carried out under acidic conditions (Scheme 2.10). When the glycerol molecule undergoes protonation the energy barrier for dehydration indeed falls from about 60 to 20 kJ/mol [65]. The reaction is preferably carried out in gas phase to limit coke formation. Yet the main difficulty of this chemistry remains the slow but steady deactivation of the catalysts due to heavy coke deposition, caused inevitably by the acidic nature of the catalysts and the high reaction temperature [66].

Several companies active in the acrylic acid market hold patents on glycerol conversion to acrolein, mostly over solid acid catalysts, from Evonik [67],

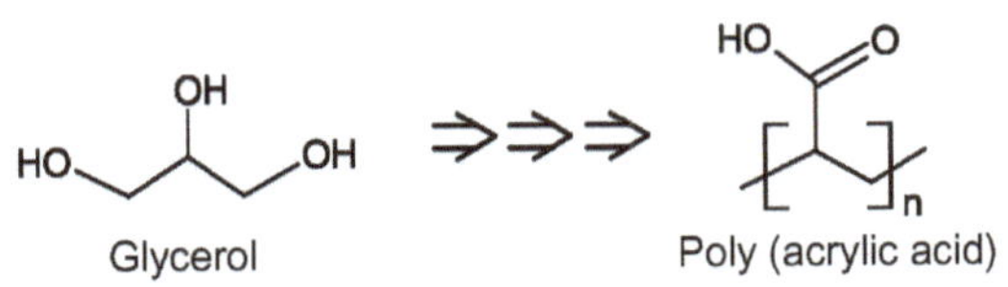

SCHEME 2.9
Glycerol is dehydrated to acrolein, which is further captively oxidized to acrylic acid. The acid is then polymerized to poly(acrylic acid).

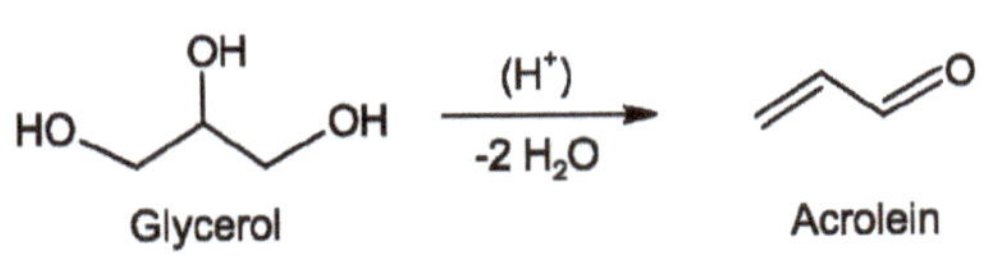

SCHEME 2.10
Acid-induced dehydration of glycerol to acrolein proceeds smoothly at 250–340°C over a heterogeneous acidic catalyst.

through Nippon Shokubai [68], Arkema [69], and Dow (Rohm and Haas) [70]. Arkema even holds a patent on the one-pot oxydehydration reaction of glycerol to acrylic acid in the presence of molecular oxygen in which dehydration is followed by aerobic oxidation of the resulting acrolein directly to acrylic acid thereby coupling the exothermic oxidation reaction (Eq. 2.3) with the endothermic dehydration reaction (Eq. 2.4) [71]:

$$CH_2OH - CHOH - CH_2OH \rightarrow CH_2 = CH - CHO + 2H_2O \quad (2.3)$$

$$CH_2 = CH - CHO + \tfrac{1}{2}O_2 \rightarrow CH_2 = CH - COOH \quad (2.4)$$

The reactor contains two catalytic beds, the first with a ground solid tungstated zirconia catalyst, and the second with a mixed W–Sr–V–Cu–Mo industrial oxidation catalyst. The gas mixture fluxes consecutively through the first and the second beds. The reactor is placed in a heated chamber at 280°C and is fed with an aqueous solution containing 20 wt% of glycerol and oxygen. The time of contact is about 2.9 s, after which all the glycerol is converted, and acrylic acid obtained in 74% yield.

Glycerol can also be dehydrated over HZSM-5 zeolite with 28 wt% acrolein yield, using 85% glycerol as feedstock, continuously regenerating the catalyst in a way similar with the one used in the fluid catalytic cracking of hydrocarbons, so that catalyst activity can be maintained over longer periods of time [72].

Assuming a $300/tonne for the glycerol feedstock, the process would translate into a $1120/tonne cost of acrolein from glycerol versus $900/tonne of acrolein from propene (assuming a $1000/tonne price for propene) [73].

The process was not commercialized. Research, though, continued, and recently hierarchically structured zeolites with open and interconnected mesopore architectures (H-Z5-3 and H-Z5-4) with thoroughly tuned Si/Al ratio to optimize acidity were shown to indeed exhibit strongly enhanced catalytic activity, stability, and selectivity compared with the conventional ZSM-5 zeolite [74].

A most promising process alternative to the tandem reaction protocol for the conversion of glycerol to acrylic acid via acrolein intermediate has been recently developed by researchers in Singapore [75]. Glycerol deoxydehydration by formic acid to allyl alcohol is followed by oxidation of the latter to acrylic acid over metal oxide catalyst with excellent yield and selectivity.

Both reactions are fast and clean. In detail glycerol heated with formic acid in final formic acid:glycerol molar ratio = 1.8:1 at 235°C under an ambient atmosphere, affords allyl alcohol in 97% yield. The formic acid is added in three portions, and the reaction is completed in 6 h with no other side product collected besides water and unreacted formic acid. Remarkably moisture and methanol that are the main potential impurities in crude bioglycerol, did not affect deoxydehydration reaction even at 20% content.

Subsequent oxidation of allyl alcohol over $Mo_8V_2WO_\delta$/SBA-15 catalyst (40% $Mo_8V_2WO_\delta$ loading) in which the catalytic centers are highly dispersed and confined within the mesopores of SBA-15 preventing sintering and agglomeration, affords full allyl alcohol conversion at temperatures above 300°C.

The catalyst supported in mesoporous SBA-15 silica shows excellent stability on time stream under the optimal reaction conditions with an overall acrylic acid yield of 80% (Fig. 2.4) and high selectivity for acrylic acid (>80%) maintained up to at least 100 h time on stream.

Thermogravimetric analysis of the used catalyst gave a <3% weight loss in the 300–600°C temperature range, pointing to negligible coke deposition even after 100 h on stream.

An alternative process based on low cost molybdenum catalysis able to convert at least 25% of the glycerol into allyl alcohol has been developed by Fristrup and coworkers [76].

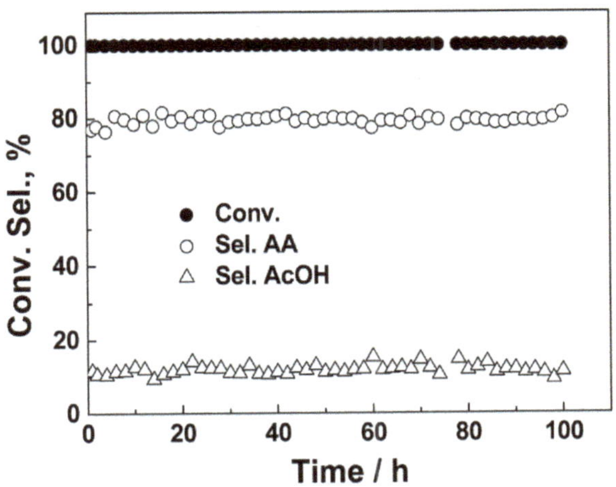

FIGURE 2.4
Time on stream test for allyl alcohol oxidation over $Mo_8V_2WO_\delta$/SBA-15; feed: 20 wt % allyl alcohol in H_2O, 0.5 mL h^{-1}; carrier gas, 20 mL min^{-1}, 10% O_2/He; reaction temperature, 340 °C. Abbreviations: AA, acrylic acid; AcOH, acetic acid; Acr, acrolein. *Reproduced from reference X. Li, Y. Zhang, Highly Efficient Process for the Conversion of Glycerol to Acrylic Acid via Gas Phase Catalytic Oxidation of an Allyl Alcohol Intermediate, ACS Catal. 2016, 6, 143–150, with kind permission.*

2.10 THE BUSINESS CASE FOR GLYCEROL-BASED ACRYLIC ACID

Every year about 350,000 tonnes of acrolein—an explosive and highly toxic [77] chemical fatal if swallowed, inhaled, and toxic if absorbed through the skin—are produced via vapor-phase oxidation of propylene in fixed-bed reactors over mixed-metal oxide containing Mo and Bi catalyst at 250–400°C with most demand originating from the manufacture of methionine, an essential amino acid needed in animal feed.

As put it by Etzkorn [78], the narrow commercial markets, coupled to the pronounced reactivity and toxicity of acrolein, have limited acrolein producers to only a few chemical companies, including Nippon Shokubai and Arkema in France.

Far larger amounts of crude acrolein are produced as an intermediate in the production of acrylic acid by the captive oxidation of acrolein to acrylic acid, a $\approx$4.5 million tonnes/year chemical used primarily as a feedstock to produce water-soluble acrylate coatings for the paint industry, and to produce superabsorbent polymers for the baby diaper industry [79].

The volatile prices of propylene, coupled with the increasing demand for acrylic acid led to a spike in the price of acrylic acid, which nearly doubled between 2009 and 2011, making attractive in the production of acrolein from readily available glycerol.

In late 2009 Nippon Shokubai announced the construction of a pilot plant in Japan using a high-performance catalyst for glycerin dehydration [80]. Similarly in 2010 Arkema started a pilot plant for converting bio-based glycerin into acrolein and acrylic acid [81].

So far, however, the glycerol dehydration route over solid acids has not been commercialized. In 2013 a reputed chemical magazine was reporting that the glycerin-based process at Arkema, a French specialty chemical company, was ready for full-scale manufacturing but, according to the manager responsible for acrylics, was put on hold due to the cost of glycerol feedstock as no "company manufacturing diapers" would be "ready to pay a 15% premium because it is a biobased raw material" because "no consumer is ready to pay a premium for the diaper" [81].

As of October 2016 the price of acrylic acid had shrinked to $900/tonne from $2300/tonne of January 2014, due both to oversupply and reduction of propylene feedstock cost.

Producing a chemical priced $900/tonne from a feedstock such as refined glycerol selling at $600/tonne simply does and will not justify investment in new technology until the conflicting dynamics between oil extractable at low cost and population growth will not determine consistently higher oil (and thus propene) price.

When such oil higher prices will be reached, acrylic acid plants using glycerol as feedstock will see the light.

2.11 BIOCHEMICAL ROUTE TO 3-HYDROXYPROPIONALDEHYDE

The other important chemical that can be produced by dehydration of glycerol is 3-HPA in which dehydration occurs via loss of a primary hydroxyl group, leading to hydroxypropionaldehyde and acrolein. 3-HPA is toxic and exhibits substantial product inhibition. In aqueous solution *Lactobacillus reuteri* exhibits significantly higher tolerance toward 3-HPA than other organisms and converts glycerol to reuterin, a natural antimicrobial, i.e., an equilibrium mixture of 3-HPA, 3-HPA hydrate, and the 3-HPA dimer (Scheme 2.11).

3-HPA is a versatile substance of significant potential interest as a chemical intermediate, because it is easily converted into a number of large-scale commodity chemicals [82]. For example, a potential large-scale use of 3-HPA is

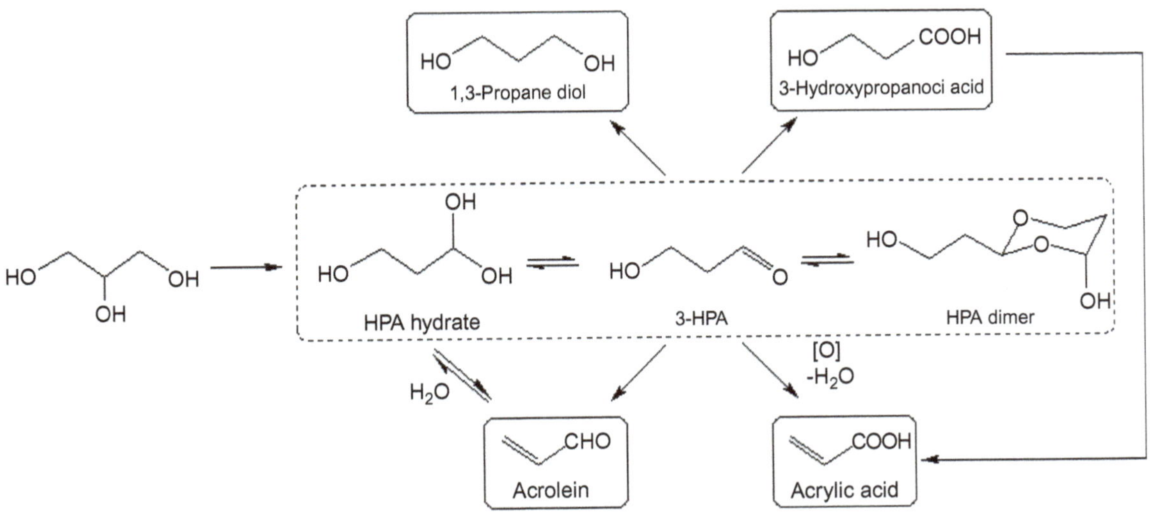

SCHEME 2.11

Production of 3-hydroxypropionaldehyde (3-HPA) and related derivatives. *Reproduced from reference J. J. Bozell, G. R. Petersen, Technology development for the production of biobased products from biorefinery carbohydrates—the US Department of Energy's "Top 10" revisited, Green Chem. 2010, 12, 539–554, with kind permission.*

in the production of PDO by combining the high yield of 3-HPA from *L. reuteri* or other processes with conventional catalytic hydrogenation [83].

In 2012 German researchers reported that the biocatalyst lifetime and product formation could be drastically increased binding the aldehyde product (toxic to cells) to carbohydrazide to form the hydrazine [84]. For the first time it was possible to recycle the immobilized biocatalyst for at least 10 cycles (overall lifetime > 33 h).

Lactobacillus reuteri cells were immobilized in polyvinyl alcohol (PVA)–based matrix that allows the entrapment of various (bio)catalysts in stable hydrogels with excellent mechanical properties (*LentiKats*) [85]. At the optimal pH value between 6.8 and 7.2 in a single fed-batch biotransformation at 45°C 138 g/L glycerol was converted into 108 g/L 3-HPA with an overall productivity of 21.6 g/L/h, namely the highest 3-HPA concentration and productivities reported until then for the microbial production of 3-HPA from glycerol.

If glucose is added to glycerol in the substrate reaction mixture, then the 3-HPA initially formed via glycerol conversion by the glycerol dehydratase enzyme is subsequently reduced by the propanediol oxidoreductase enzyme (NADH) eventually affording PDO [86].

A further increase in PDO productivity (QPDO) with *L. reuteri* was lately achieved by researchers in Canada and Brazil who carried out the fermentation using glucose and glycerol as cosubstrates with the genus *Lactobacillus* under different modes of reactor operation [87]. In continuous mode, excellent 0.70 g_{PDO}/g_{GLY} yield was achieved and productivity in the steady state was 20% higher (4.92 g/L/h), compared to the best result in the repeated batch mode. The theoretical mass conversion from glycerol to PDO is 0.83 g_{PDO}/g_{GLY}. The transformation of glycerol into PDO is the energetically most favorable pathway for this microorganism to regenerate the NADH produced in glucose catabolism. Accordingly the steady-state condition of the continuous mode favors PDO production, because it allows the highest rate of glucose consumption. Besides PDO the process also affords lactate with high conversion yield, namely providing the monomer needed for the production of biodegradable polylactic (PLA) polymers produced from renewable resources [88].

2.12 POLYHYDROXYALKONOATES

In early 2016 Italy's company Bio-on announced a partnership with an Italian sugar company for the construction the world's first facility for the production of polyhydroxyalkanoates (PHAs, Fig. 2.5), bioplastic from biodiesel glycerol

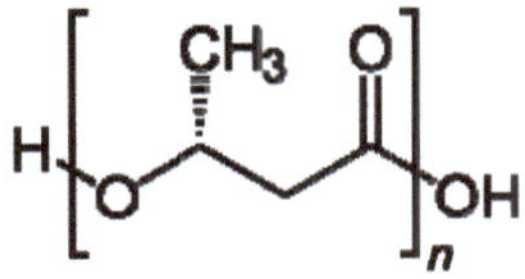

FIGURE 2.5
Structure of poly-(R)-3-hydroxybutyrate (P3HB), a polyhydroxyalkanoate (PHA).

at a new 5000 tonnes/year plant (expandable to 10,000 tonnes/year) [89]; while other plants were being authorized and constructed by the end of 2016 in France and Brazil, licensed to use the Bio-On technology.

PHAs are thermoplastic polyesters with many of the same physical properties of polyethylene (PE), polypropylene (PP), synthetic polyesters, and acrylics, but the PHA-based plastic is biodegradable and biocompatible. Current uses for PHAs include packaging, disposable syringes and bottles, fast food containers, and coatings.

The company licensed from the University of the Hawaii biotechnology developed in the early 2000s by Yu and coworkers to produce bio-based PHA materials from renewable feedstocks, such as agricultural and food processing waste by-products.

The technology uses two bioreactors coupled by a membrane to produce PHA. In the first reactor, the raw food waste is anaerobically fermented to more bioavailable substrates such as lactic, butyric, acetic, and propionic acids.

The acids are then continuously moved through the membrane into a second reactor by molecular diffusion, wherein a culture of the bacteria *Ralstonia eutropha* use the acids as a carbon source to make PHA. At the end of fermentation process, the bacterial cells can accumulate 60%–70% biopolyester of their mass [90].

Yu's team also developed in 2006 a solvent-free technology to recover the PHA biopolymers from cell mass with high yield (>95 wt%) using sequential treatment of the product with aqueous solutions of mineral acid, base, and a small amount of oxidation agent for decolorization purpose [91]. No waste is produced from this process, and the remaining solid mass from fermentation can be sold as soil conditioner.

The bioplastics technology consists of three parts, including pretreatment of feedstocks into suitable substrates for a special type of microbial organism, high cell density fermentation for biosynthesis of biopolyesters, and solvent-free recovery and purification of biopolyesters to make the final product of bioplastics.

The company improved the process for recovering and purifying PHAs from a cell culture [92], by inserting a tangential filtration step of the PHA suspension to obtain a concentrated suspension that, following bleaching, is mildly dried obtaining the PHA granules ready for commercialization.

One new application of the PHA thereby obtained is a new formulation, called *Minerv PHA Bio Cosmetics* [93], suitable for replacing plastic microbeads widely used by the cosmetic industry in products such as "scrub" products, providing the exfoliating effect for deep cleansing, lipstick and lip gloss, mascara, eye-liner, nail polish, shampoo, foam bath, and toothpaste.

Conventional microbeads made of polymers such as PE, polymethyl methacrylate (PMMA), nylon, polyethylene terephthalate (PET) and PP, pollute the environment because once entering the environment plankton in rivers and seas swallow but cannot metabolize these microscopic plastic particles and thus introduce them into the food chain.

The level of pollution is so serious that the US government recently enacted new legislation (Microbead-Free Waters Act of 2015) [94] banning rinse-off cosmetics that contain intentionally-added plastic microbeads beginning on January 1, 2018, and to ban manufacturing of these cosmetics beginning on July 1, 2017.

The use of *Minerv PHA Bio Cosmetics* in cosmetics eliminates these pollutants because the PHA microparticles actually decompose into nutrients for some microorganisms and plants present in nature and in general quickly biodegrade in water not entering the food chain. Furthermore the new PHA-based formulation is capable of binding active molecules and antioxidants, such as Coenzyme Q10, vitamins, proteins, and similar active substances present in cosmetic products, transporting them to parts of the body where those products are normally applied.

We were writing in the second edition of *The Future of Glycerol* (2010):

> Human chemical ingenuity has rapidly created a route to glycerol derivatives that find application in fields as diverse as chemicals and fuels, and the automotive, pharmaceutical, detergent and building industries...
>
> It is now markets, rather than ideas, which generate products [95].

The forthcoming PHA production from bioglycerol may be one nice example of this forecast, as "markets" (the Italy's based entrepreneurs) acted to identify a valued idea not yet commercialized, to eventually develop a series of advanced products whose value is much higher than conventional use of bioplastics to make bags, syringes, cups, or packaging stuff.

2.13 GLYCEROL OXIDATION BIOPRODUCTS

Glycerol oxygenate derivatives obtained via selective oxidation of one, two, or three hydroxyl groups of the glycerol are valued fine chemicals for which new chemical, electrochemical, and biological catalytic routes have been lately developed [96].

Reviewing said advances in 2011 with Kimura, Dumeignil and coworkers were noting how oxidation of glycerol had not been commercially applied yet [96]. They ascribed this finding to the fact that dihydroxyacetone (DHA) was actually the only compound with a significant market as sunless tanner.

The same team calculated the production costs for DHA, tartronic acid (TA) and mesoxalic acid (MA) obtained via aerobic oxidation over noble metal catalysts and concluded that only the synthesis of DHA was commercially feasible, with the catalyst price still accounting for 95% of the production costs even after 10 times of reuse, due the presence of expensive Pt in the noble metal catalysts used for the synthesis of TA and MA. Since then many new catalysts have been introduced, which justify commercial production of at least some of the oxidation products shown in Fig. 2.6.

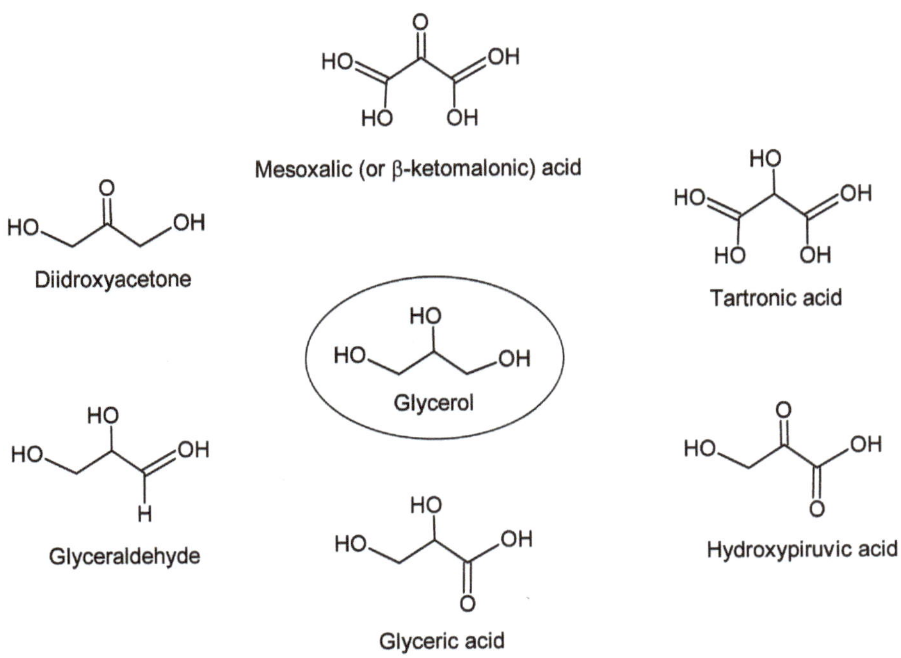

FIGURE 2.6
Products of glycerol selective oxidation.

Dihydroxyacetone: As mentioned earlier, dihydroxyacetone (DHA) is the main active ingredient in artificial tanning agents [97], even though DHA-induced melanoidins in skin act as a topical sunscreen attenuate the formation of healthy vitamin D [98]. The market of several thousand tonnes per year grows at >10% yearly rate with sunless tanners revenues now exceeding $800 million [99].

DHA is industrially produced via microbial conversion of glycerol via *Gluconobacter oxydans* with high selectivity to DHA but with low productivity and high production cost [100]. Modern bioreactors adopt a semicontinuous process in which part of the fermentation broth is allowed to remain in the reactor as the inoculate for the next cycle thereby avoiding tedious inoculation required for each cycle [101].

A major progress occurred in 2013 when Xu and coworkers in China discovered a green, high-yield process driven by visible light to convert glycerol dissolved in water into DHA over nanostructured bismuth wolframate using oxygen as primary oxidant [102]. Photoactivity test on reuse of Bi_2WO_6 for six operational runs showed no significant loss of activity and selectivity toward oxidation of glycerol to DHA in water under visible light irradiation (Fig. 2.7).

Even better, the process can be run over Bi_2WO_6 sol–gel entrapped inside a transparent porous silica matrix thereby enhancing glycerol adsorption, thus improving its local concentration and favoring reaction with the photogenerated active species [103], affording DHA at low cost with no waste

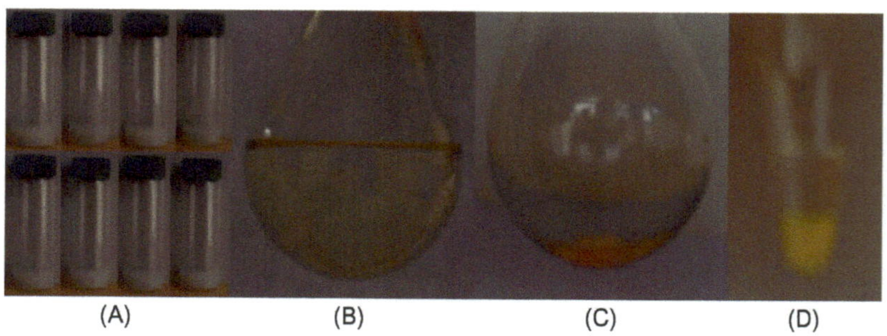

FIGURE 2.7

(A) Pictures of the suspension after the photocatalytic conversion of glycerol in water over under visible light irradiation for 2 h; (B) the suspension after removing the catalyst particles via centrifugation; (C) the remaining solution after removing water solvent via a rotary evaporation; (D) the two-layered solution in a 1.5-mL centrifugal tube after centrifugation. The upper layer solution is the remaining glycerol. The lower layer with yellow color is DHA. *Reproduced from reference Y. Zhang, N. Zhang, Z.-R. Tang, Y.-J. Xu, Identification of Bi2WO6 as a highly selective visible-light photocatalyst toward oxidation of glycerol to dihydroxyacetone in water, Chem. Sci. 2013, 4, 1820–1824, with kind permission.*

generation in water solvent under ambient conditions with molecular oxygen as oxidant and visible light as the driving energy source.

Recently a team in China led by Fu has shown that hydroxymethylfurfural (4-HMF), a precursor of several functional molecules for both pharmaceutical and material sciences, can be obtained from DHA in high yield through consecutive base-catalyzed condensation and acid-catalyzed dehydration steps [104].

Remarkably to demonstrate the practical potential of the process in vista of industrial manufacture, the team performed the two-step conversion in continuous process, proving the prolonged stability (36 h) of catalytic activity of the Amberlite basic and Amberlyst acid resins.

In the process (Fig. 2.8), an aqueous solution of 10% DHA first passes through a fixed-bed reactor (R1) packed with basic resin. The resulting dendroketose solution S1 was condensed continuously under reduced pressure. Subsequently a DMSO solution of 10% dendroketose passes through the

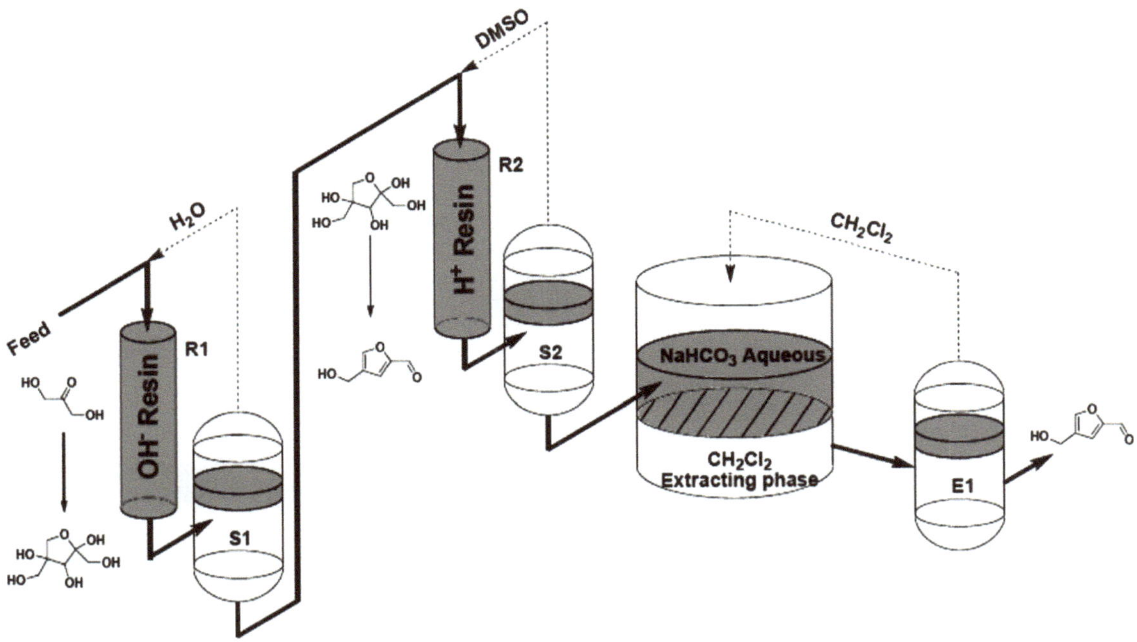

FIGURE 2.8

The process for conversion of DHA to hydroxymethylfurfural (HMF). Aldol condensation of DHA to form dendroketose in a fixed-bed reactor over a basic resin (R1); evaporation of water (S1); dehydration of dendroketose to 4-HMF over a H^+ resin (R2); evaporation of DMSO from the solvent containing 4-HMF (S2); and separation of HMF from the extracting solvent (E1). *Reproduced from reference M.-S. Cui, J. Deng, X.-L. Li, Y. Fu, Production of 4-hydroxymethylfurfural from derivatives of biomass-derived glycerol for chemicals and polymers, ACS Sustainable Chem. Eng. 2016, 4, 1707–1714, with kind permission.*

second fixed-bed reactor (R2) packed with acidic resin affording 4-HMF in a solution that was condensed under reduced pressure prior to be added into $NaHCO_3$ aqueous solution from which 4-HMF continuously extracted with dichloromethane was eventually obtained in 80% isolated yield.

References

[1] R. Auvergne, S. Caillol, G. David, B. Boutevin, J.-P. Pascault, Biobased thermosetting epoxy: present and future, Chem. Rev. 114 (2014) 1082–1115.

[2] D. Witter, IBISWorld Industry Report 32521 Plastic & Resin Manufacturing in the US, IBISWorld, 2016.

[3] Solvay, Sustainable bio-based ech for improving the environmental impact of products through epoxy resins, InnoBioPlast 2016, 15–16 September 2016, Bangkok, Thailand.

[4] Grand View Research, Epoxy Resin Market Analysis by Application (Paints & Coatings, Wind Turbine, Composites, Construction, Electrical & Electronics, Adhesives) and Segment Forecasts to 2024, San Francisco, 2016.

[5] R. Tesser, E. Santacesaria, M. Di Serio, G. Di Nuzzi, V. Fiandra, Kinetics of glycerol chlorination with hydrochloric acid: a new route to α,γ-dichlorohydrin, Ind. Eng. Chem. Res. 46 (2007) 6456–6465.

[6] P. Krafft, P. Gilbeau, B. Gosselin, S. Claessens, Process for producing dichloropropanol from glycerol, the glycerol coming eventually from the conversion of animal fats in the manufacture of biodiesel, WO 2005054167 (2005).

[7] D. Siano, E. Santacesaria, V. Fiandra, R. Tesser, G. Nuzzi, M. Di Serio, et al., Process for the production of alpha, gamma-dichlorohydrin from glycerin and hydrochloric acid, WO 2006111810 A2 (2006).

[8] A. Almena, M. Martín, Technoeconomic analysis of the production of epichlorohydrin from glycerol, Ind. Eng. Chem. Res. 55 (2016) 3226–3238.

[9] F. Vleugels, Spolchemie, Global Petrochemicals 7th Annual Meeting—World Refining Association, Vienna, February 9–11, 2010.

[10] International EPD Consortium, Epichlorohydrin G, EPD Registration Number: S-EP-00283, 30 April 2011.

[11] SCS Global Services Report, Certification Evaluation Report, SCS Certificate Code-SCS-RSB/PC-0018, September 2015.

[12] T. Caulier, Epicerol and the environmental challenges of bio-based chemistry, 6th Workshop on Fats and Oils as Renewable Feedstock for the Chemical Industry, Karlsruhe, March 17–19, 2013.

[13] J.E. Oude Lenferink, P. Tavares, B. Sierra, Vintage, the new cool, Opportunities for Europe's Chemical Industry?, Industrial Technologies 2016, Amsterdam, June, 22–24, 2016.

[14] D. de Guzman, Solvay bags award for bio-ECH, greenchemicalsblog.com, October 26, 2015.

[15] M. Chong, China's Yihai Kerry to restart Lianyungang ECH plant on 20 August, icis.com, January 20, 2011.

[16] Solvay China Communication, 2016 Solvay in China, solvay.cn.

[17] J.-F. Tremblay, Dow selects Shanghai for epichlorohydrin, Chemical & Engineering News, March 29, 2007.

[18] B.M. Bell, J.R. Briggs, R.M. Campbell, S.M. Chambers, P.D. Gaarenstroom, J.G. Hippler, et al., Glycerin as a renewable feedstock for epichlorohydrin production: the GTE process, Clean 36 (2008) 657–661.

[19] D. de Guzman, Growing glycerine-to-ECH plants, icis.com, January 20, 2011.

[20] D. de Guzman, Glycerol use: the next generation, 2nd ICIS Pan American Oleochemicals Conference, October 11–12, 2016 Miami, FL.

[21] Petrosil, Crude Glycerine FOB Malaysia prices decrease by 5% since start of the year, glycerinereport.com, October 2016.

[22] Jim Lane, Biofuels mandates around the world: 2016, biofuelsdigest.com January 3, 2016.

[23] D.J. Mutaqin, After B20 policy, biomass is our energy future, The Jakarta Post, January 7, 2016.

[24] E. Chow, Malaysia in talks to move startup of B10 biodiesel program to July, reuters.com, June 20, 2016.

[25] N. Notman, BPA: friend or foe?, ChemistryWorld, November 20, 2012.

[26] M. Fache, B. Boutevin, S. Caillol, Epoxy thermosets from model mixtures of the lignin-to-vanillin process, Green Chem. 18 (2016) 712–725.

[27] E.D. Hernandez, A.W. Bassett, J.M. Sadler, J.J. La Scala, J.F. Stanzione III, Synthesis and characterization of bio-based epoxy resins derived from vanillyl alcohol, ACS Sustainable Chem. Eng. 4 (2016) 4328–4339.

[28] O. Faruk, M. Sain, Lignin in Polymer Composites, Elsevier, Boston, 2015.

[29] N. Smolarski, High Value Opportunities for Lignin Unlocking its Potential Market Insights, Frost and Sullivan Market Research, London, 2012.

[30] M. Fache, B. Boutevin, S. Caillol, Vanillin production from lignin and its use as a renewable chemical, ACS Sustainable Chem. Eng. 4 (2016) 35–46.

[31] E.A. Baroncini, S. Kumar Yadav, G.R. Palmese, J.F. Stanzione III, Synthesis and characterization of bio-based epoxy resins derived from vanillyl alcohol, Appl. polym 133 (2016) 44103.

[32] T. Iwata, Biodegradable and bio-based polymers: future prospects of eco-friendly plastics, Angew. Chem. Int. Ed. 54 (2015) 3210–3215.

[33] F. Meneguzzo, R. Ciriminna, L. Albanese, M. Pagliaro, The great solar boom: a global perspective into the far reaching impact of an unexpected energy revolution, Energy Sci. Eng. 3 (2015) 499–509.

[34] E.S. Vasiliadou, A.A. Lemonidou, Glycerol transformation to value added C3 diols: reaction mechanism, kinetic, and engineering aspects, WIRE Energy Environ 4 (2015) 486–520.

[35] Merchant Research & Consulting, World propylene glycol market to reach supply-demand balance in 2015, Birminghan, April 18, 2014.

[36] M. Chong, Asia MPG outlook stable to soft as supply expected to lengthen, icis.com, July 19, 2016.

[37] U.S. Food and Drug Administration, CFR—Code of Federal Regulations Title 21, Part 184—Direct Food Substances Affirmed as Generally Recognised as Safe, Sec. 184.1666 Propylene glycol.

[38] T.Y. Lim, R.L. Poole, N.M. Pageler, Propylene glycol toxicity in children, J. Pediatr. Pharmacol. Ther. 19 (2014) 277–282.

[39] Ethylene glycol is gravely toxic to humans and animals by oral exposure, and when ingested, depressing the central nervous system, can be fatal even in small quantities. See D. Jacobsen, K.E. McMartin, Methanol and ethylene glycol poisonings. Mechanism of toxicity, clinical course, diagnosis and treatment, Med. Toxicol 1 (1986) 309–334.

[40] TechSci Research, Global Deicing Fluid Market Forecast & Opportunities, 2021, New York, 2016.

[41] Y. Xi, J.E. Holladay, J.G. Frye, A.A. Oberg, J.E. Jackson, D.J. Miller, A kinetic and mass transfer model for glycerol hydrogenolysis in a trickle-bed reactor, Org. Process Res. Dev. 14 (2010) 1304–1312.

[42] J. Henkelmann, M. Becker, J. Bürkle, P. Wahl, G. Theis, S. Maurer, Process for the preparation of 1,2-propanediol, WO2007099161, (2007).

[43] P. Poetschacher, Extending oleochemistry: glycerin to bio propylene glycol, 2nd ICIS Pan American Oleochemicals Conference, Miami, October 11–12, 2016.

[44] M.R. Nanda, Z. Yuan, W. Qin, C. Xu, Recent advancements in catalytic conversion of glycerol into propylene glycol: a review, Catal. Rev 58 (2016) 309–336.

[45] M.W.M. Tuck, S.N. Tilley, Process, WO2007010299 (2007).

[46] http://davyprotech.com/what-we-do/licensed-processes-and-core-technologies/licensed-processes/propylene-glycol/specification/ (Last time accessed 22.10.2016).

[47] H. Finch, Profile: Europe monopropylene glycol, icis.com, 19 April 2013.

[48] Oleon, Radia 7207, oleonhealthandbeauty.com.

[49] B. Balmer, PTT—A new alternative for fibres and textiles? frost.com, June 23, 2004.

[50] K. Thanh Lam, J.B. Powell, P.R. Weider, Process for preparing 1,3-propanediol, WO1997016250 (1997).

[51] M. Emptage, S.L. Haynie, L.A. Laffend, J.P. Pucci, G. Whited, Process for the biological production of 1,3-propanediol with high titer, US7067300 (2000).

[52] D. de Guzman, 2015: A turning point for bioplastic building-block investments, Plastics Engineering, March 2016, p. 19.

[53] J.V. Kurian, A new polymer platform for the future—sorona from corn derived 1,3-propanediol, J. Polym. Environ. 13 (2005) 159–167.

[54] R. Lin, H. Liu, J. Hao, K. Cheng, D. Liu, Enhancement of 1,3-propanediol production by *Klebsiella pneumoniae* with fumarate addition, Biotechnol. Lett. 27 (2005) 1755–1759.

[55] R. Lin, H. Liu, J. Hao, K. Cheng, D. Liu, 1,3-Propanediol and its copolymers: research, development and industrialization, Biotechnol. J. 5 (2010) 1137–1148.

[56] D. de Guzman, Bio-1,3 PDO market update, greenchemicalsblog.com, March 18, 2013.

[57] C.C.M., Glory, Biomaterial's first phase of biological PDO programme enters pilot production, kcomber.com, March 18, 2013.

[58] S. Hirschmann, K. Baganz, I. Koschik, K.-D. Vorlop, Development of an integrated bioconversion process for the production of 1,3-propanediol from raw glycerol waters, Landbauforsch Völkenrode 55 (2005) 261–267.

[59] M. Gungormusler-Yilmaz, N. Cicek, D.B. Levin, N. Azbar, Cell immobilization for microbial production of 1,3-propanediol, Crit. Rev. Biotechnol. 36 (2016) 482–494.

[60] METabolic EXplorer, announces launch of construction work on its first PDO manufacturing plant in Malaysia, Clermont-Ferrand, November 2, 2010.

[61] F. Berthon, Metabolic explorer: Nous ne repartons pas de zéro, loin de là ! tradingsat.com, December 18, 2015.

[62] B. Katryniok, S. Paul, V. Bellière-Baca, P. Rey, F. Dumeignil, Glycerol dehydration to acrolein in the context of new uses of glycerol, Green Chem. 12 (2010) 2079–2098.

[63] L. Liu, X.P. Ye, J.J. Bozell, A comparative review of petroleum-based and bio-based acrolein production, ChemSus. Chem. 5 (2012) 1162–1180.

[64] A. Pavone, Acrylic acid from glycerin, Process Economics Program, Report No. 6E, SRI Consulting, Menlo Park: 2011.

[65] M.R. Nimlos, S.J. Blanksby, X. Qian, M.E. Himmel, D.K. Johnson, Mechanisms of glycerol dehydration, J. Phys. Chem. A 110 (2006) 6145–6156.

[66] M. Dalil, M. Edake, C. Sudeau, J.-L. Dubois, G.S. Patience, Coke promoters improve acrolein selectivity in the gas-phase dehydration of glycerol to acrolein, Appl. Catal. A: Gen 522 (2016) 80–89.

[67] A. Neher, T. Haas, D. Arntz, H. Klenk, W. Girke, Process for production of acrolein and other oxygenated compounds from glycerol in a transported bed reactor, US5387720 (1995).

[68] E. Matsunami, T. Takahashi, H. Kasuga, Y. Arita, Process for production of acrolein, WO 2007119528 (2007).

[69] Y. Magatani, K. Okumura, J.-L. Dubois, J.-F. Devaux, Catalyst and process for preparing acrolein and/or acrylic acid by dehydration reaction of glycerin, US9162954B2 (2015).

[70] L.E. Bogan, Jr., M.A. Silvano, Process for production of acrolein from glycerol, US8198477B2 (2012).

[71] J.-L. Dubois, C. Duquenne, W. Hölderich, Method for producing acrylic acid from glycerol, US7910771 (2006).

[72] P. O'connor, C.A. Corma, G. Huber, L.A. Savanaud, Process for production of acrolein and other oxygenated compounds from glycerol in a transported bed reactor, WO2008052993 (2008).

[73] A. Corma, G. Huber, L. Sauvanaud, P. O'connor, Biomass to chemicals: catalytic conversion of glycerol/water mixtures into acrolein, reaction network, J. Catal. 257 (2008) 163–171.

[74] H. Zhang, Z. Hu, L. Huang, H. Zhang, K. Song, L. Wang, et al., Dehydration of glycerol to acrolein over hierarchical ZSM-5 zeolites: effects of mesoporosity and acidity, ACS Catal 5 (2015) 2548–2558.

[75] X. Li, Y. Zhang, Highly efficient process for the conversion of glycerol to acrylic acid via gas phase catalytic oxidation of an allyl alcohol intermediate, ACS Catal 6 (2016) 143–150.

[76] J.R. Dethlefsen, P. Fristrup, Process for functionalizing biomass using molybdenum catalysts WO 2015/028028 A1 (2014).

[77] A. Moghe, Molecular mechanisms of acrolein toxicity: relevance to human disease, Toxicol. Sci. 143 (2015) 242–255.

[78] W.G. Etzkorn, Acrolein and derivatives, Kirk-Othmer Encyclopedia of Chemical Technology, John Wiley & Sons, New York, 2009, pp. 1–29.

[79] Markets and Markets, Acrylic acid market by derivative (Esters/Acrylates—METHYL, ETHYL, BUTYL, 2-EH; Polymers—Elastomers, Superabsorbent Polymers, Water Treatment Polymers; Other Derivatives), by applications & region—Global Forecast to 2020, Pune, 2016.

[80] NIPPON SHOKUBAI announces developing of a technology for acrylic acid production from biomass resources, shokubai.co.jp, October 26, 2009.

[81] Cit. in A.H. Tullo, Hunting for biobased acrylic acid, Chem. Eng. News 91 (2013) 18–19.

[82] J.J. Bozell, G.R. Petersen, Technology development for the production of biobased products from biorefinery carbohydrates—the US Department of Energy's "Top 10" revisited, Green Chem 12 (2010) 539–554.

[83] S. Vollenweider, C. Lacroix, 3-hydroxypropionaldehyde: applications and perspectives of biotechnological production, Appl. Microbiol. Biotechnol 64 (2004) 16–27.

[84] H. Krauter, T. Willke, K.-D. Vorlop, Production of high amounts of 3-hydroxypropionaldehyde from glycerol by *Lactobacillus reuteri* with strongly increased biocatalyst lifetime and productivity, New Biotechnol 29 (2012) 211–217.

[85] GeniaLabm Technology LentiKats, lentikats.eu

[86] T.L. Talarico, L.T. Axelsson, J. Novotny, M. Fiuzat, W.J. Dobrogosz, Utilization of glycerol as a hydrogen acceptor by *Lactobacillus reuteri*: purification of 1,3-propanediol:NAD oxidoreductase, Appl. Environ. Microbiol. 56 (1990) 943–948.

[87] P.B. Vieira, B.V. Kilikian, R.V. Bastos, E.A. Perpetuo, C.A.O. Nascimento, Process strategies for enhanced production of 1,3-propanediol by Lactobacillus reuteri using glycerol as a co-substrate, , Biochem. Engineer. J 94 (2015) 30–38.

[88] R. Kozlovskiy, V. Shvets, A. Kuznetsov, Technological aspects of the production of biodegradable polymers and other chemicals from renewable sources using lactic acid, *J. Clean. Prod.* DOI: 10.1016/j.jclepro.2016.08.092

[89] Bio-Based World News, Italy claims world's first facility to produce bio-plastic from glycerol, biobasedworldnews.com, January 4, 2016.

[90] G. Du, J. Yu, Green technology for conversion of food scraps to biodegradable thermoplastic polyhydroxyalkanoates, Environ. Sci. Technol. 36 (2002) 5511–5516.

[91] J. Yu, L.X. Chen, Cost-effective recovery and purification of polyhydroxyalkanoates by selective dissolution of cell mass, Biotechnol. Prog. 22 (2006) 547–553.

[92] S. Begotti, Process for recovering and purifying polyhydroxyalkanoates from a cell culture, US 20160168319 A1 (2016).

[93] M. Astorri, New formulations based on revolutionary PHA bioplastic, Future of Formulations in Cosmetics, Budapest, May 18, 2016.

[94] Public Law No: 114-114, Microbead-Free Waters Act of 2015, December 28, 2015.

[95] M. Pagliaro, M. Rossi, The Future of Glycerol, Second ed., RSC Publishing, Cambridge, 2010, p. 161.

[96] B. Katryniok, H. Kimura, E. Skrzynska, J.-S. Giradon, P. Fongarland, M. Capron, et al., Selective catalytic oxidation of glycerol: perspectives for high value chemicals, Green Chem. 13 (2011) 1960–1979.

[97] R.K. Chaudhuri, Dihydroxyacetone: chemistry and applications, in: M.L. Schlossman (Ed.), third ed., Self-Tanning Products, Chemistry and Manufacture of Cosmetics, vol. 3, C.H.I.P. S.Books, Weimar, Texas, 2002.

[98] L.A.G. Armas, R.M. Fusaro, R.M. Sayre, C.J. Huerter, R.P. Heaney, Do melanoidins induced by topical 9% dihydroxyacetone sunless tanning spray inhibit vitamin d production? A pilot study, Photochem. Photobiol 85 (2009) 1265–1266.

[99] E. Brooke, The self-tanning industry is still booming, fashionista.com, April 15, 2014.

[100] Z. Zheng, M. Luo, J. Yu, J. Wang, J. Ji, Novel process for 1,3-dihydroxyacetone production from glycerol. 1. Technological feasibility study and process design, Ind. Eng. Chem. Res 51 (2012) 3715–3721.

[101] R. Bauer, N. Katsikis, S. Varga, D. Hekmat, Study of the inhibitory effect of the product dihydroxyacetone on *Gluconobacter oxydans* in a semi-continuous two-stage repeated-fed-batch process, Bioprocess Biosyst. Eng. 28 (2005) 37.

[102] Y. Zhang, N. Zhang, Z.-R. Tang, Y.-J. Xu, Identification of Bi_2WO_6 as a highly selective visible-light photocatalyst toward oxidation of glycerol to dihydroxyacetone in water, Chem. Sci 4 (2013) 1820–1824.

[103] Y. Zhang, R. Ciriminna, G. Palmisano, Y.-J. Xu, M. Pagliaro, Sol-gel entrapped visible light photocatalysts for selective conversions, RSC Adv 4 (2014) 18341–18346.

[104] M.-S. Cui, J. Deng, X.-L. Li, Y. Fu, Production of 4-hydroxymethylfurfural from derivatives of biomass-derived glycerol for chemicals and polymers, ACS Sustainable Chem. Eng. 4 (2016) 1707–1714.

[105] H. Habe, T. Fukuoka, D. Kitamoto, K. Sakaki, Biotechnological production of d-glyceric acid and its application, Appl. Microbiol. Biotechnol. 84 (2009) 445.

[106] T. Fukuoka, H. Habe, D. Kitamoto, K. Sakaki, Bioprocessing of glycerol into glyceric acid for use in bioplastic monomer, J. Oleo Sci 60 (2011) 369–373.

[107] T. Nikaido, Y. Kobayashi, Manufacture of d-glyceric acid with gluconobacter species, Jpn. Kokai Tokkyo Koho JP (1989) 011–68292.

[108] S. Carrettin, P. McMorn, P. Johnston, K. Griffin, G.J. Hutchings, Selective oxidation of glycerol to glyceric acid using a gold catalyst in aqueous sodium hydroxide, Chem. Commun. (2002) 696–697.

[109] H. Tan, O.E. Tall, Z. Liu, N. Wei, T. Yapici, T. Zhan, et al., Selective oxidation of glycerol to glyceric acid in base-free aqueous solution at room temperature catalyzed by platinum supported on carbon activated with potassium hydroxide, ChemCatChem 8 (2016) 1699–1707.

[110] G. Caselli, M. Mantovanini, C.A. Gandolfi, M. Allegretti, S. Fiorentino, L. Pellegrini, et al., Generation of drugs affecting bone metabolism, J. Bone Miner. Res. 12 (1997) 972–981.

[111] T. Power, Oxygen scavenger for low moisture environment and methods of using the same, WO2005040304 (2005).

[112] X. Tiang, Z. Wang, P. Yang, R. Hao, S. Jia, Na Li, et al., Green oxidation of bio-lactic acid with H_2O_2 into tartronic acid under UV irradiation, RSC Adv 6 (2016) 41007–41010.

[113] J. Cai, H. Ma, J. Zhang, Z. Du, Y. Huang, J. Gao, et al., Catalytic oxidation of glycerol to tartronic acid over Au/HY catalyst under mild conditions, J. Catal. 35 (2014) 1653–1660.

CHAPTER 3

Esters, Ethers, Polyglycerols, and Polyesters

3.1 GLYCEROL ESTERS

Generally obtained via chemical glycerolysis or enzymatic transesterification, glycerol esters of fatty acids are amphiphilic molecules useful as nonionic surfactants and emulsifiers. In general esterification of glycerol with carboxylic acids results in monoacylglycerols (MAGs), diacylglycerol (DAG), and triacylglycerol (TAG). All esters have a wide variety of commercial applications. Both MAGs and DAGs are widely used as food additives in bakery products, margarines, dairy products, and sauces.

In the cosmetic industry they are employed as texturing agents for improving the consistency of creams and lotions. DAGs are used as a plasticizer and softening agent and solvent. TAGs are employed as solvent, antimicrobial, and emulsifying agents in cigarette filters and pharmaceuticals.

Noting that the core technology for fatty ester production was mature and thoroughly established, in 2004 analysts at Frost & Sullivan emphasized that new opportunity existed for synthesizing specialty esters due to increasing sophistication of product formulations to meet the increasingly stringent regulatory environment, particularly in Europe, and the increasing popularity of naturally-derived products [1].

In the subsequent years this insight showed its validity. Glycerol monostearate (GMS; Fig. 3.1) has become the largest product segment of the 1.14 million tonnes global fatty acid ester market in 2014, accounting for 39.3% [2].

Growing at >4% annual rate, demand for GMS is driven by its use as a food thickener, emulsifying agent for oils, waxes, and solvents, protective coating for hygroscopic powders, solidifier and control release agent in pharmaceuticals, and resin lubricant.

Diacetin (diacetylglycerol, E1517) and triacetin (triacetalglycerol, E1518) are approved in Europe as solvents in the flavoring industry [3], and

Glycerol. DOI: http://dx.doi.org/10.1016/B978-0-12-812205-1.00003-5

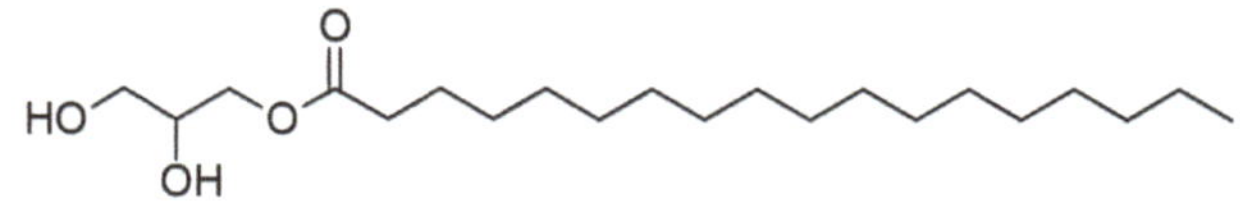

FIGURE 3.1
Chemical structure of glycerol monostearate (GMS).

in the United States by the Food and Drug Administration as GRAS (generally recognized as safe) substances. Triacetin is used as a food additive, both as a flavoring solvent and a humectant for the solvency of flavorings, as additive in cigarettes as humectant.

Tributyrin, the glyceryl ester of butanoic acid, has anticancer properties and is more potent than natural butyrate [4].

By the same token, esterification of valeric acid with glycerol produces mono, di and trivalerin. The latter triester shows good compatibility with waterborne polymer dispersions, improving film formation properties which make them excellent as coalescing agents in waterborne paint, adhesive and sealant [5]. Both monobutyrin and trivalerin used as feed additives for chickens reduces *Salmonella enteritidis* colonization [6].

Glycerol esters are traditionally manufactured industrially by continuous chemical glycerolysis of fats and oils at high temperature (220–250°C), employing alkaline catalysis under a nitrogen atmosphere [7]. Manufacture involves heating a stirred emulsion of vegetable oil and glycerol in the presence of a strongly basic inorganic catalyst such as KOH, NaOH, or $Ca(OH)_2$ affording MAG and DAG (Eqs. 3.1 and 3.2, respectively):

$$\text{Triglyceride} + 2\text{glycerol} \rightarrow 3\text{Monoglyceride} \tag{3.1}$$

$$2\text{Triglyceride} + \text{glycerol} \rightarrow 3\text{Diglyceride} \tag{3.2}$$

The yield of MAG is usually around 40%, and the crude product, colored due to thermal degradation products, is distilled to give a food-grade material (90% MAG).

DAG has two isomers namely, 1,2-DAG and 1,3-DAG, which undergo acyl migration to form equilibrium at a ratio of 3-4:7-6 between 1,2- and 1,3-DAG, as 1,3-DAG is more thermodynamically stable because of steric effects [8]. The 1,2-DAG in edible oil is largely converted to the 1,3-DAG by migration of the acyl group during high temperature processing [9].

State-of-the-art production of 1,3-diacylglycerols uses lipase-catalyzed transesterification under solvent-free conditions. For example, Novozym 435 and Lipozyme RM IM (*Rhizomucor miehei*) afford 52% and 60.7% DAG at 32 h, respectively (Fig. 3.2) [10]. Easily found throughout the

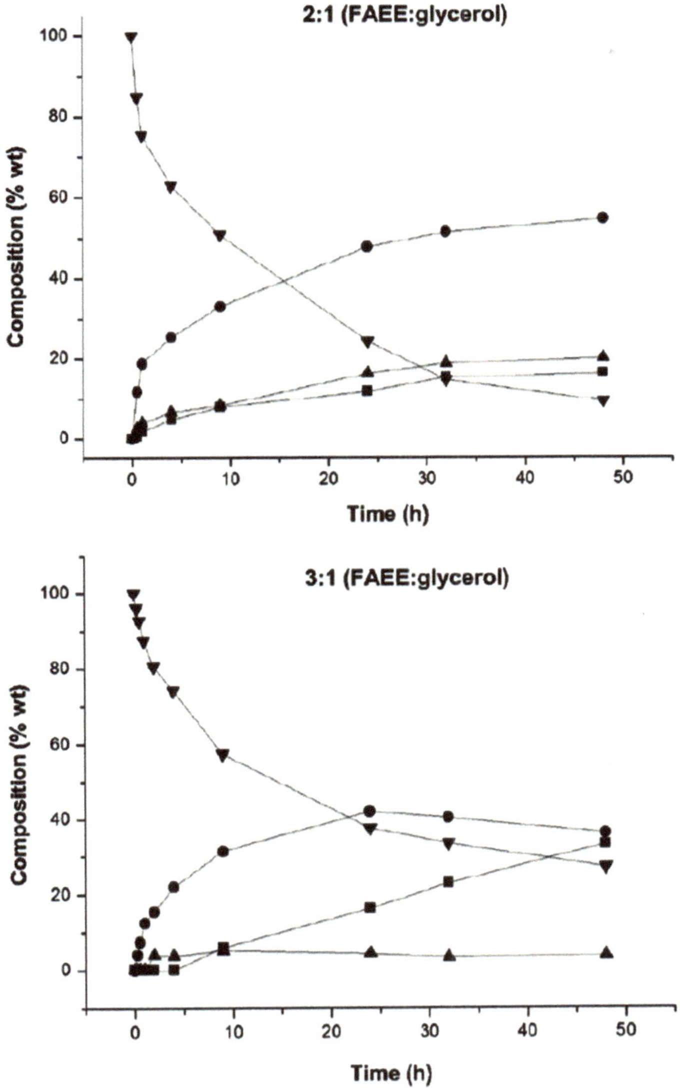

FIGURE 3.2

Composition (% w/w) of fatty acid ethyl esters (FAEE) (▼), TAG (■), DAG (●), and MAG (▲) at different reaction times of transesterification between FAEE and raw glycerol with Novozym 435 (*Candida antarctica*). Substrate molar ratio 2:1 (FAEE:glycerol): 4 g FAEE and 1 g glycerol. Substrate molar ratio 3:1 (FAEE:glycerol): 6.5 g FAEE and 1 g glycerol. *Reproduced from reference L. Vázquez, N. González, G. Reglero, C. Torres, Solvent-Free Lipase-Catalyzed Synthesis of Diacylgycerols as Low-Calorie Food Ingredients, Front Bioeng Biotechnol. 2016, 4:6, with kind permission.*

animal and plant kingdoms, as well as in molds and bacteria, lipases (triacylglycerol acylhydrolase, EC 3.1.1.3) are hydrolases of exceptional versatility widely used by industry in the food, detergent, and pharmaceutical sectors. The enzymes act on carboxylic ester bonds, to hydrolyze triglycerides into diglycerides, monoglycerides, fatty acids, and glycerol; as well as catalyze esterification and transesterification reactions in nonaqueous media.

The use of lipases in place of inorganic acid or base to catalyze the solvent-free glycerolysis of fats and oils has several advantages: catalysis at lower temperatures indeed prevents the discoloration and alteration of unsaturated fatty acids that is common at elevated temperatures, with state-of-the-art production involving immobilized lipases such as sol–gel entrapped lipases employed in industry for the synthesis of MAGs and DAGs via continuous glycerolysis over immobilized enzyme packed reactors [11].

Comparison of the traditional and enzymatic processes for the production of MAG shows the key advantages of the bioprocess in terms of higher selectivity (and less waste), and lower temperature and pressure. In general already in 2005 Martinez and coworkers in Spain could conclude that the enzyme cost for small-scale production need to drop to around €100/kg, whereas for large-scale a cost around a few 10 €/kg was needed [12].

DAG has the advantage of being stable to decomposition at cooking temperatures. A DAG cooking oil was produced by a Japanese corporation from soybean and canola oil using enzymatic glycerolysis with lipase and marketed under the tradename "Healthy Econa Oil" for use in cooking, frying, and dressings between 1998 and 2009.

Research concerning the human nutritional characteristics of DAG oil in which 1,3-DAG is the major component compared to TAG oil had shown a significant suppressive effect of DAG on body fat accumulation, due to the reduced possibility of synthesis of TAG in the small intestine following DAG oil digestion and thus to reduced body weight and visceral fat mass [13].

Shortly after the introduction to the market this oil became the best-selling cooking oil in Japan. In 2004, a new joint venture started to manufacture and market the oil with the Evona brand in the Americas, Europe, Australia and New Zealand.

In the same year the manufacturing company submitted a request to the European Food Safety Authority (EFSA) to market Enova oil as a food ingredient for fat spreads/margarines, dressings for salads/mayonnaise, bakery products, yoghurt, health bars, and health drinks. An initial assessment by the Dutch Competent Authority reached the conclusion that it was safe for human consumption.

Some of the other member states of the EU raised concerns and objections but with the exception of Spain, these were satisfied by further safety data supplied by the manufacturing company. The European Commission thus asked EFSA to provide a scientific opinion on the use of Enova oil as a food ingredient in view of the concerns raised, that included the possible use of genetically modified materials, details of the production process, the stability of the product, the *trans* fatty acid content, and other possible adverse health effects on potentially sensitive groups. The EFSA concluded that the product is safe for human consumption but that in order for it not to be nutritionally disadvantageous to consumers, the *trans* fatty acid content should be reduced to the level in the conventional vegetable oils that the novel oil is intended to replace [14].

However, in December 2007 the German Federal Institute for Risk Management (BfR) was mandated by the Ministry of Food, Agriculture and Consumer Protection to validate an analytical method for the determination of 3-chloropropane-1,2-diol (3-MCPD) fatty acid esters in refined vegetable oils. In March 2009 the BfR published the results of the assessment of glycidol esters in several common refined vegetable oils, particularly palm oil [15]. Estimate of potential dietary intake, especially from margarine and commercial dairy products for infants, the genotoxic and carcinogenic properties for glycidol, and the likelihood for hydrolytic metabolism to bioavailable glycidol, led the BfR to conclude that current levels of exposure of infants and some adults could present a hazard to human health. The BfR recommended that the levels of glycidol esters in vegetable oils should be reduced as far as possible.

In response to these findings, the Japanese corporation conducted analysis on Econa Cooking Oil and found that the amount of glycidol ester in Econa Cooking Oil was 91 ppm, namely from 200 to 100 times higher than 0.5–9.1 ppm levels in common food oils [16]. In September thus the company announced temporary suspension of the production and sale of Econa products, including sales of Evona oil in North America, with other companies following in South Korea and elsewhere [17].

Production and sale, however, did not restart; and in 2016 a market report of Canada's Government on functional foods and beverages in Japan was reporting that "the oils and fats category saw the most significant decrease from 2010 to 2015, with a 20% decline in the value of retail sales. This drop may be due to Kao Corporation's decision to halt sales of their Econa brand products in 2009 because of possible concerns regarding the safety of glycidol fatty acid ester" [18].

3.2 BIOPOLYOL-BASED POLYURETHANES

A promising route to bio-based polyurethane (PU) foams of high performance based on biopolyols derived from crude biodiesel glycerol has been developed by Li and coworkers in the United States [19]. In detail the team has discovered that biopolyols produced from crude glycerol using one-pot thermochemical conversion process without the addition of extra catalysts and reagents are ideally suited to afford high-quality PU foams.

Crude glycerol is simply heated to 110°C for 1 h followed by heating to 150°C for 2.5 h or 5 h under reduced pressure (0.001 MPa). At 150°C dehydration of glycerol to glycidol or acrolein does not occur, whereas the reduced pressure effectively removes water and methanol formed during esterification and transesterification of glycerol with the free fatty acids (FFAs) present in the crude glycerol mixture, which indeed are almost entirely consumed after 5 h.

The resulting biopolyols (polyol-2.5h and polyol-5h) mainly comprised of monoglycerides and diglycerides with a bi-hydroxyl structure with branched long fatty acid ester chains (Scheme 3.1), are then reacted with commercial polymeric methylene-4,4′-diphenyl diisocyanate (pMDI).

The procedure used to produce the PU foams used standard conditions, mixing vigorously the biopolyol with silicone surfactant, catalyst, and water to achieve an homogenous dispersion, to which pMDI was added pouring the resulting mixture in a cylinder to grow at ambient temperature and then cure overnight. A foam obtained by reacting crude glycerol was also prepared (PU-CG).

Table 3.1 shows that the thermal conductivity of PU-5h (38 mW/m/K) is comparable to the 33 mW/m/K thermal conductivity of the Daltofoam TE 44204 foam commercialized by Huntsman (a leading PU manufacturer).

The microstructure of the resulting foams present largely distinct morphologies (Fig. 3.3). Glycerol, a short-chain extender with three hydroxyls, contributes to the formation of a crosslink structure in which stiff polymer chains are formed resulting in a liquid membrane lacking elasticity, which cannot sustain bubbles during the foam formation process.

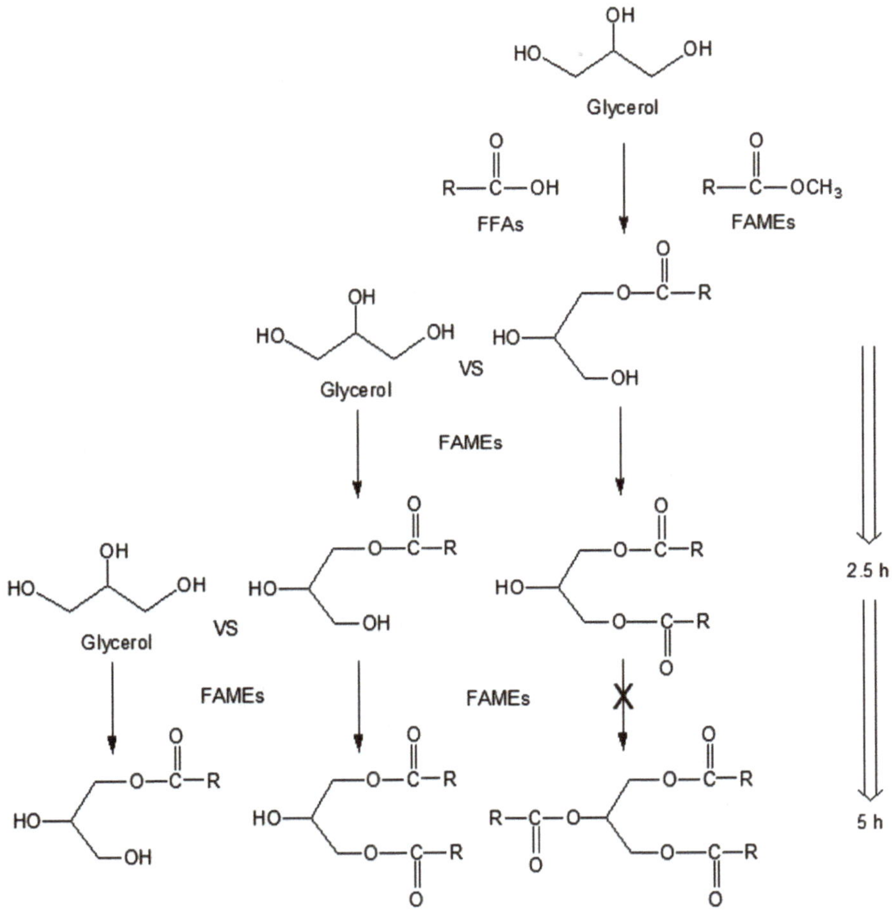

SCHEME 3.1

Schematic diagram of major reaction process during biopolyol synthesis. *FFA*, free fatty acid; *FAME*, fatty acid methyl ester. *Reproduced from reference C. Li, X. Luo, T. Li, X. Tong, Y. Li, Polyurethane foams based on crude glycerol-derived biopolyols: one-pot preparation of biopolyols with branched fatty acid ester chains and its effects on foam formation and properties, Polymer 55, (2014) 6529–6538, with kind permission.*

On the other hand, the structure of the biopolyols predominant in polyol-5h, with its branched fatty acid ester chains, mixes with urea-based segments in the PU, weakening microphase separation, helping to sustain foaming bubbles, and reducing bubble merging during foam formation, eventually affording a PU foam with small cell size, which translates into thermal conductivities comparable to commercial foams made from petroleum-based polyols. A company based in Ohio, Bio100 Technologies, was formed in 2009 to develop and commercialize the eco-friendly polyol technology under an exclusive global license with The Ohio State University.

Table 3.1 Physical Properties of polyurethane (PU) Foams

Property	PU-CG	PU-2.5h	PU-5h
Thermal conductivity mW/(m K)	52	41	38
Density (kg/m^3)	77.3	65.7	58.3
Compressive strength (kPa)	110	93	82
Modulus of compression (MPa)	1.4	2.1	2.1

Reproduced from reference C. Li, X. Luo, T. Li, X. Tong, Y. Li, Polyurethane foams based on crude glycerol-derived biopolyols: One-pot preparation of biopolyols with branched fatty acid ester chains and its effects on foam formation and properties, Polymer 55 (2014) 6529–6538, with kind permission.

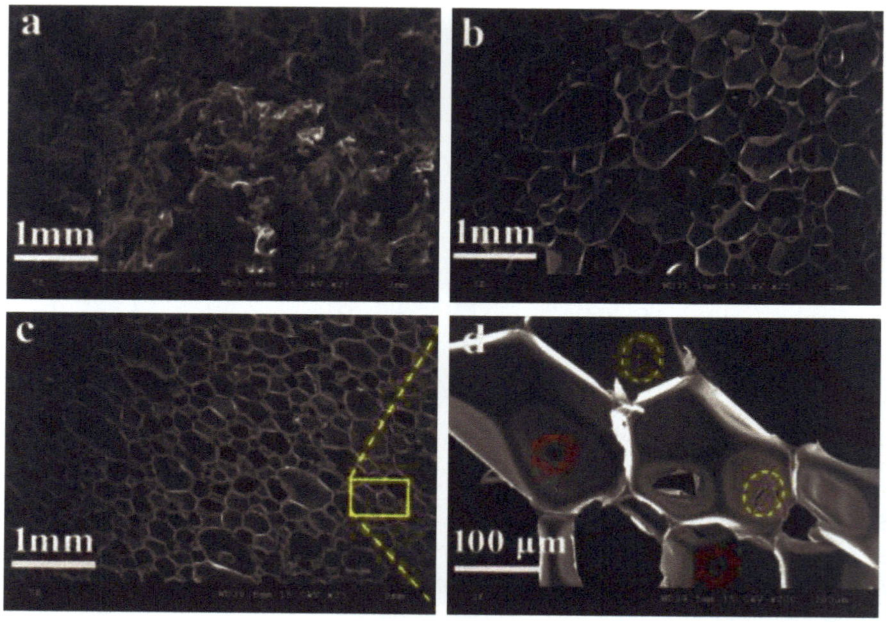

FIGURE 3.3

SEM pictures of PU-CG (A), PU-2.5h (B), PU-5h (C), and PU-5h (D) at higher magnification. *Reproduced from reference C. Li, X. Luo, T. Li, X. Tong, Y. Li, Polyurethane foams based on crude glycerol-derived biopolyols: one-pot preparation of biopolyols with branched fatty acid ester chains and its effects on foam formation and properties, Polymer 55, (2014) 6529–6538, with kind permission.*

3.3 GLYCEROL CARBONATE

Glycerol carbonate (4-hydroxymethyl-1,3-dioxolan-2-one; GC) is one of the most attractive value-added derivatives of glycerol [20]. The carbonate has a high boiling point (351°C) and a low melting point (−66.7°C). Its dipolar

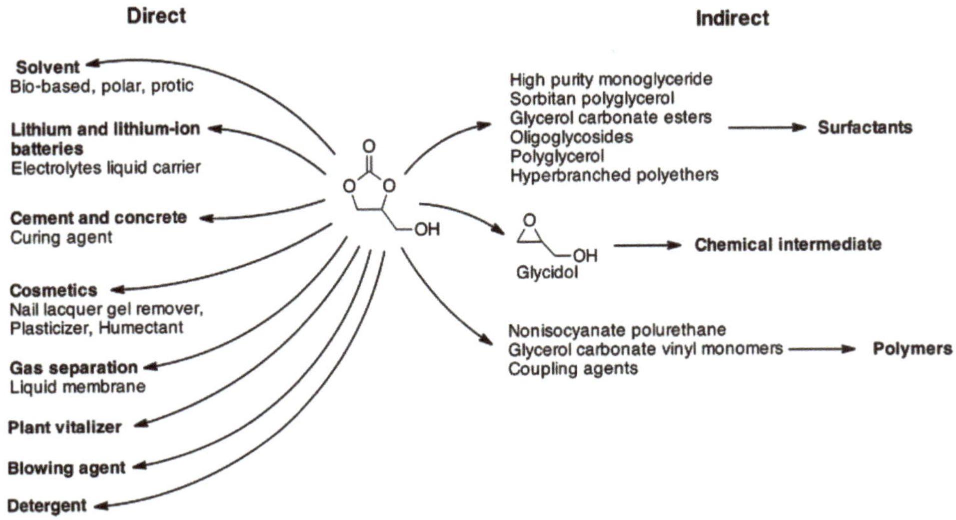

SCHEME 3.2

Direct and indirect applications of glycerol carbonate (GC). *Reproduced from reference M.O. Sonnati, S. Amigoni, E.P. Taffin de Givenchy, T. Darmanin, O. Choulet, F. Guittard, Glycerol carbonate as a versatile building block for tomorrow: synthesis, reactivity, properties and applications, Green Chem. 15 (2013) 283–306, with kind permission.*

moment ($\mu = 5.4$ D) and dielectric constant ($\varepsilon = 109.7$) are higher than most other organic compounds. Being also nontoxic, once available at low-cost GC will be used as low-volatile organic compound (VOC) general purpose polar solvent for plastics and resins such as cellulose acetate, nylon, nitrocellulose, and polyacrylonitrile, an electrolyte liquid carrier or an additive in lithium and lithium-ion batteries, a curing agent in cement and concrete, an additive in cosmetics, a liquid–gas separation system, a detergent, a plant vitalizer, and a blowing agent.

Furthermore GC has great potential as a versatile building block due to the wide reactivity associated to the concomitant presence of a hydroxyl group and a 2-*oxo*-1,3-dioxolane group, enabling its conversion into chemical intermediates suitable for the production of epichlorohydrin, surfactants, and polymers (Scheme 3.2).

The price of GC, reported in 2014 to exceed \$8100/tonne, has limited so far commercial applications to only a few thousand tonnes per year [21].

Comparing in 2012 the alternatives for industrial manufacturing of CC, Ochoa-Gómez and coworkers concluded that indirect synthetic routes starting from glycerol and CO_2-derivatives, rather than from glycerol and CO_2, were the most attractive [22].

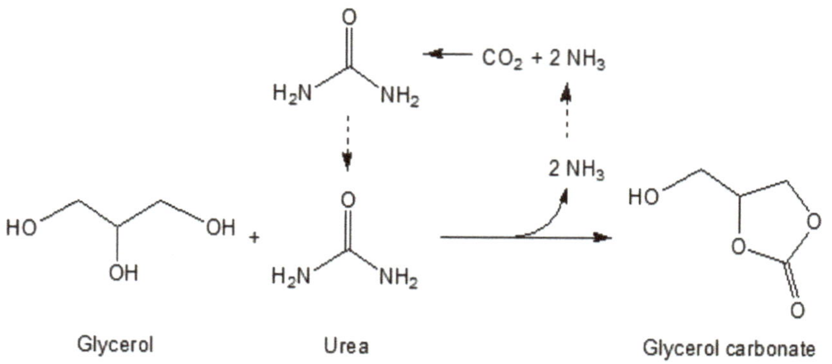

SCHEME 3.3
Glycerolysis with urea, uses urea as CO_2 donor.

One of the most attractive has been developed in the late 1990s by Mouloungui and coworkers in France who discovered a method affording high yields (>95%) of GC by catalytic reaction between equimolar amounts of urea and glycerol. The reaction is catalyzed by zinc sulfate at temperature around 150°C. Reduced pressure (40 mbar) shifts the equilibrium of the reaction toward GC product by eliminating ammonia in the gaseous phase (Scheme 3.3) [23].

This glycerolysis process using urea as CO_2 donor is very attractive as it uses two readily available raw materials in a chemical cycle that, overall, results in the chemical fixation of CO_2. Furthermore closing the cycle, ammonia formed as by-product can be converted again into urea using CO_2 captured from the atmosphere.

The process can be catalyzed by solid base [24] or by solid acids. Reaction over Zn-modified H-ZSM-5 zeolite affords up to 80% glycerol conversion and 100% selectivity to GC (much higher than with homogenous zinc sulfate), with remarkably good catalyst stability and resistance to deactivation by impurities in the glycerol feed [25].

Reviewing the synthesis and applications of GC in 2013, Guittard and coworkers were noting that "the first industrial productions of GC seem to be based on the transcarbonation of glycerol with organic carbonate sources such as dimethyl carbonate or propylene carbonate using basic catalysts" [20].

On the other hand, in the same year Nguyen and Demirel compared the economics of GC manufacturing via glycerol direct carboxylation and glycerolysis using basic catalysis assuming a selling price for GC, identical for both plants, of $2400/tonne [26].

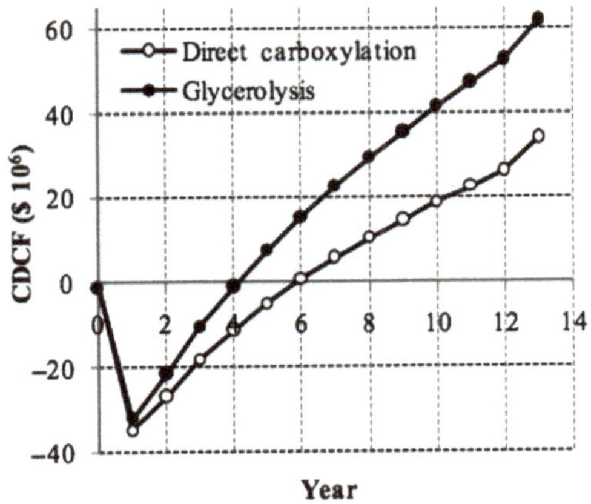

FIGURE 3.4
Comparison of the cumulative discounted cash flow (CDCF) diagrams of the direct carboxylation and glycerolysis routes. *Reproduced from reference N. Nguyen, Y. Demirel, Economic analysis of biodiesel and glycerol carbonate production plant by glycerolysis, J. Sustain. Bioen. Syst. 3 (2013) 209–216, with kind permission.*

The outcomes of the comparative analyses were clear: due to the low yield (less than 35%) of the thermodynamically limited direct carboxylation of glycerol and CO_2, the payback period of the glycerolysis plant is 2.4 years versus 3.7 years for the direct carboxylation plant; while the cash flow rate of return (Fig. 3.4) would be 32.08% for the glycerolysis plant and 19.91% for the direct carboxylation factory. The base-catalyzed glycerolysis process indeed affords 89% GC yield (with 98.6% selectivity) at 140°C, 3 kPa, with a glycerol:urea mass ratio of 3.07 over 1% La_2O_3 stable catalyst.

A further progress to economic viability was achieved 2 years later by researchers in Malaysia who discovered that the catalytic carbonylation of glycerol with urea is effectively catalyzed by gypsum ($CaSO_4 \cdot 2H_2O$) obtained as waste (and thus available at no or even negative cost) from a large mineral ore industrial processing plant [27].

The catalytic activity and selectivity are due to the concomitant action of Ca^{2+} as Lewis acid sites, which activate the carbonyl group of urea, and $(SO_4)^{2-}$ as conjugate base. activating hydroxyl group on glycerol to form GC [28].

Truly heterogeneous catalysis is observed with gypsum calcined in air for 3 h at 800°C, which affords an insoluble γ-$CaSO_4$ phase with a similar catalytic performance in several consecutive reaction cycles (73.8% GC yield and

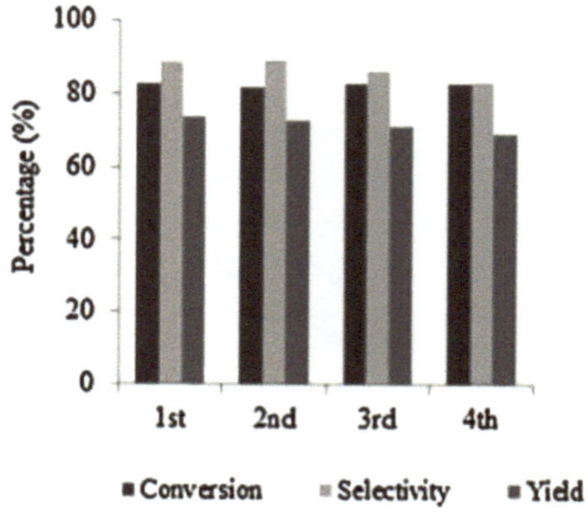

FIGURE 3.5

Reusability study of γ-CaSO4 (Gyp800) gypsum catalyst on glycerol carbonylation with urea. Reaction temperature, 150°C. *Reproduced from reference N. A. S. Zuhaimi, V. P. Indran, M. A. Deraman, N. F. Mudrikah, G. P. Maniam, Y. H. Taufiq-Yap, M. H. Ab. Rahim, Reusable gypsum based catalyst for synthesis of glycerol carbonate from glycerol and urea, Appl. Catal. A 502 (2015) 312–319, with kind permission.*

89.1% glycerol conversion). The catalyst, furthermore, consistently affords high yields of GC in several consecutive runs (Fig. 3.5), with negligible leaching of calcium.

A recent finding of great practical relevance in the surfactant industry is that surfactants obtained from GC–glycerol oligomeric esters are better than widely used surfactants, such as esters of ethoxylated sorbitans, polyethylene monoethers, and polyglycerol (PG) esters [29].

In detail GC was oligomerized in the presence of glycerol affording amphiphilic polyhydroxylated oligomers ($M_w < 1000$ Da) rich in linear carbonate groups in which the polar function is exerted by glycerol and GC rather than ethylene oxide as in most commercial surfactants. Partial ester are then produced by reaction of said oligo-(GC–glycerol) with copra oil in the presence of sodium methylate at 142°C under reduced pressure (40 mbar) for 8 h, using equimolar amounts of OH groups in the oligomer and acyl groups in the oil to generate the desired lauric ester.

The resulting linear polyhydroxylated GC–glycerol oligomeric skeleton partially functionalized by pendant fatty acids (Fig. 3.6) provides the amphiphilic structure rich in hydroxyl, carbonate, and ether functions promoting high interfacial activity leading to particularly low critical micelle concentration (CMC).

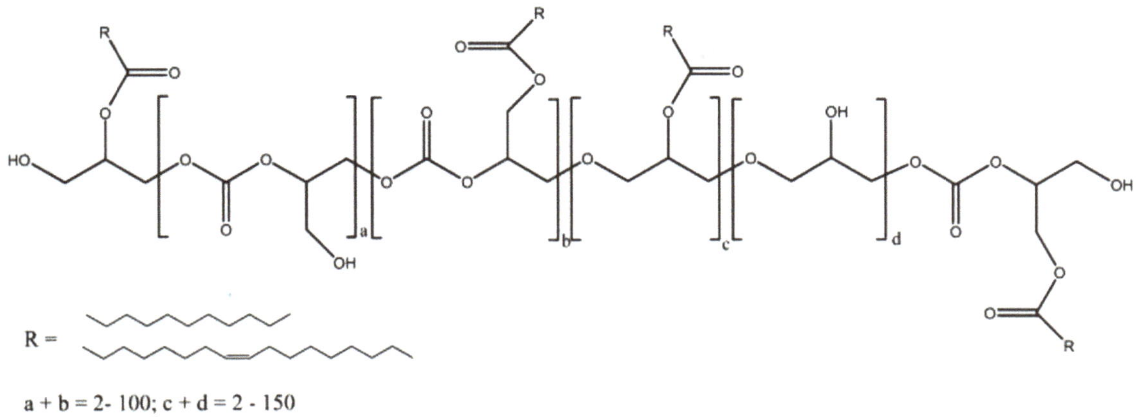

FIGURE 3.6

Structure of esters of oligo-(glycerol carbonate—glycerol ether). *Reproduced from reference S. Holmiere, R. Valentin, P. Marechal, Z. Mouloungui, Esters of oligo-(glycerol carbonate-glycerol): New biobased oligomeric surfactants, J. Coll. Interf. Sci. 487 (2017) 418—425, with kind permission.*

Table 3.2 Comparison of the Physicochemical Parameters of Ethoxylated Surfactants, Glycerol-Based Surfactants and Partial and Total Esters of Oligo-(Glycerol Carbonate—Glycerol Ether)

Surfactant	CMC (mg/L)	γ(mN/m)
Tween 80	15.7	–
Tween 20	73.7	38
$C_{12}EO_4$	23.5	35.2
$C_{12}EO_8$	37.7	41
$C_{12}EO_6$	31.9	38.5
3GML	78.9	31.4
4GML	118	35.2
5GML	168.2	39.6
Lauric oligoester OHV = 325	0.75	32.8
Lauric oligoester OHV = 223	2.2	26

CMC, *critical micelle concentration;* OHV = *hydroxyl value mg KOH/g.*
Adapted from reference S. Holmiere, R. Valentin, P. Marechal, Z. Mouloungui, Esters of oligo-(glycerol carbonate-glycerol): New biobased oligomeric surfactants, J. Coll. Interf. Sci. 487 (2017) 418—425, with kind permission.

The CMC values of the total and partial esters of oligocarbonates are very low (Table 3.2), below those of commercial ethoxylated surfactants containing oleic acid (Tween 80) or lauric acid as the lipophilic moiety (Tween 20, dodecyl polyethylene monoethers, PG monolaurates), as well those of glycerol monolaurates.

Furthermore the surface tension reached at the CMC with the new oligocarbonate esters are similar or lower than those of ethoxylated surfactants, enabling to achieve the same decrease in surface tension with smaller amounts of oligocarbonate esters than ethoxylated surfactants.

3.4 GLYCEROL ETHERS

Glycerol ethers are alkyl ethers, namely they belong to a class of molecules on which >1200 research articles and >2000 patents have been, respectively, published and granted in the 2000–14 decade [30]. Their synthesis can be achieved in many ways, out of which acid-catalyzed etherification of glycerol with alkenes or alcohols, and Pd-catalyzed telomerization (linear dimerization of 1,3-dienes with simultaneous addition of a nucleophile) are the green methods suitable for large-scale applications [31].

The catalytic reaction of glycerol with isobutene was first investigated in 1992 by Behr and Lohr at Henkel aimed at forming glycerol tertiary butyl ethers [31]. The reaction proceeds in the presence of an acid catalyst at temperatures from 50°C to 150°C and at molar ratios of glycerol:isobutylene of 1:2 or higher. The team discovered active homogeneous and heterogeneous catalysts, for instance *p*-toluene sulfonic acid, acidic ion exchangers like Amberlyst 15 and several synthetic zeolites. Sulfonic-modified mesostructured silicas are even better catalysts [32], affording no isobutylene oligomerization products.

A simple, low-cost solvent-free etherification of glycerol with long-chain alcohols affording nonionic surfactants via one-pot heterogeneous interfacial acidic catalysts has been lately demonstrated by Feng and coworkers at a Solvay/CNRS joint laboratory in Shangai (Scheme 3.4) [33].

The catalyst is an amphiphilic random sulfonated polystyrene sample (PSt-PSSA) bearing the synergetic effect of emulsifier and catalyst. Grafted on silica, the catalyst was still active and more easily separated from the reaction mixture, though the dodecanol conversion decreased from 61% to 49% after two recycling (while keeping almost the same selectivity).

In detail, the team was able to convert glycerol and dodecanol in 60%–71% yield with limited production of didodecyl ether (DE) affording alkylpolyglycerylether (AGEM) with >80% selectivity, using a 1:4 molar ratio of dodecanol to glycerol, under 200 mbar reduced pressure for 24 h at 150°C under a nitrogen atmosphere.

The resulting mixture of monolauryl polyglyceryl ethers, multilauryl polyglyceryl ethers, and multilauryl cyclicpolyglyceryl ethers (Fig. 3.7) was tested on cotton, polyester, and polyester–cotton based on 11 standard stains

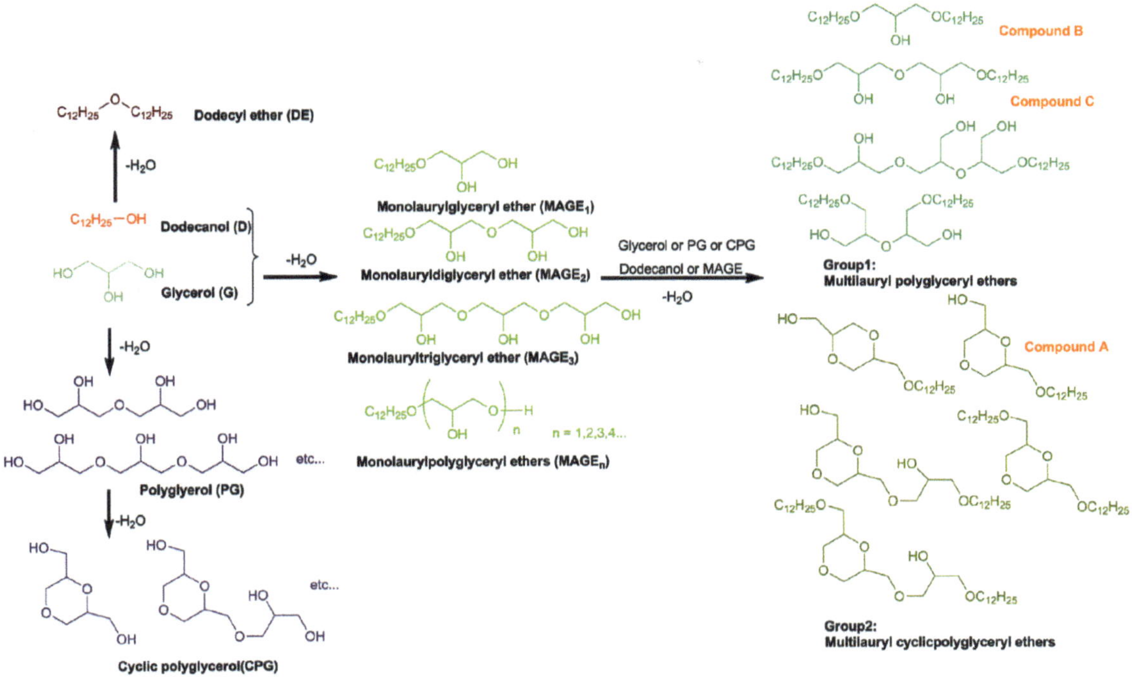

SCHEME 3.4

Representative structures of possible components inside dodecanol and glycerol etherification mixtures. *Reproduced from reference Z. Fan, Y. Zhao, F. Preda, J.-M. Clacens, H. Shi, L. Wang, et al., Preparation of bio-based surfactants from glycerol and dodecanol by direct etherification, Green Chem. 17 (2015) 882–892, with kind permission.*

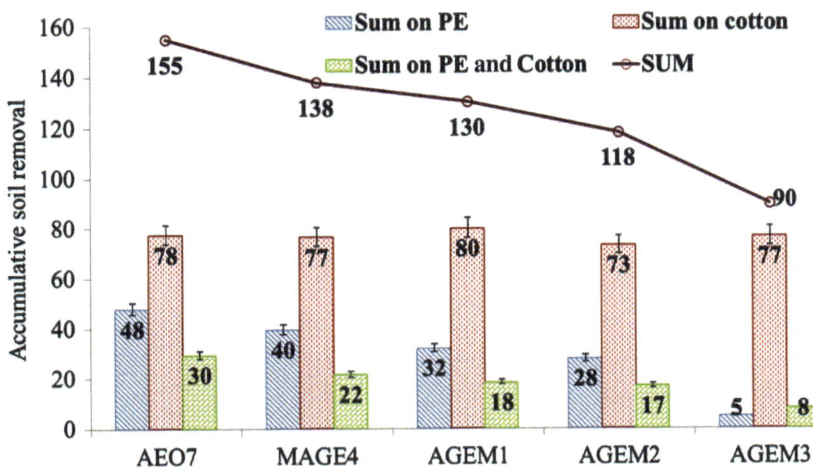

FIGURE 3.7

Laundry performance on polyester (PE), cotton, and polyester and cotton (65/35). The stains removal was adjusted based on active content. Polyester and cotton: mixed fabric with 65% polyester and 35% cotton. *Reproduced from reference Z. Fan, Y. Zhao, F. Preda, J.-M. Clacens, H. Shi, L. Wang, et al., Preparation of bio-based surfactants from glycerol and dodecanol by direct etherification, Green Chem. 17 (2015) 882–892, with kind permission.*

and compared to the performance of AEO7 (laurylpolyethoxylate-7 ether) and MAGE4 (monoalkylpolyglyceryl ether) commercial surfactants as benchmarks.

The accumulative stain removal on cotton, the increasingly chosen fabric today, for AGEM1 was better than AEO7 and MAGE4 although the performance on polyester was a little worse. Overall AGEM samples showed slightly lower accumulative stain removal ratio compared to AEO7 and MAGE4 as benchmarks. The team was expecting to prepare AGEM samples with better performance via the adjustable synthetic strategy.

Glycerol ether surfactants are preferred over glycerol esters due to higher stability of the ether compared to the ester bond.

Another interesting route to C8-chain mono-, di-, and triethers of glycerol suitable for the production of surfactant or detergent molecules is the direct telomerization of either pure or crude glycerol with 1,3-butadiene mediated by a palladium-based molecular catalyst discovered by Weckhuysen and coworkers in 2008 (Fig. 3.8) [34].

Concerning the product distribution, Pd/TPPTS (TPPTS = trisodium salt of tri(*m*-sulfonylphenyl)phosphine) results mainly in monoether formation,

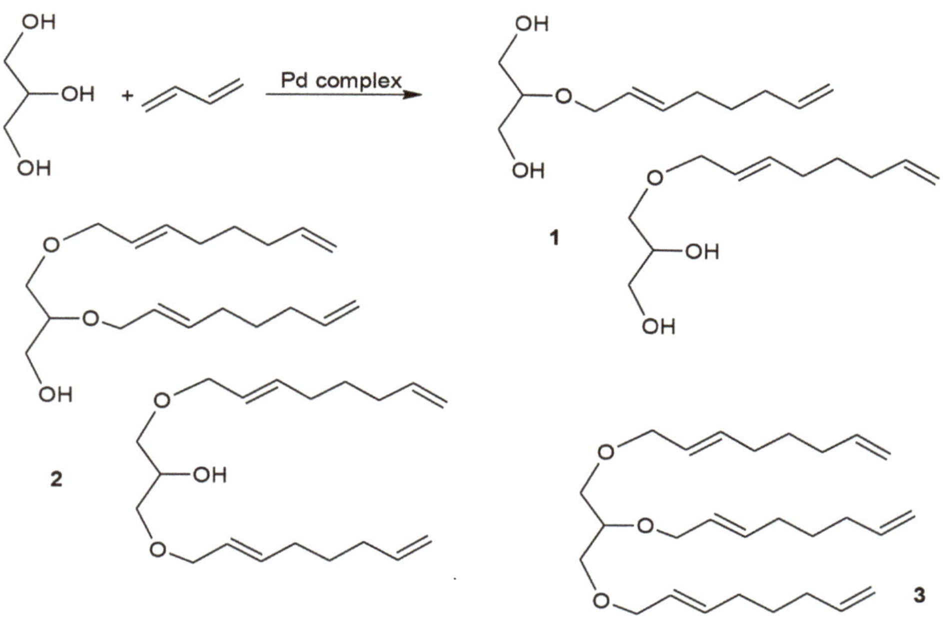

FIGURE 3.8

Telomerization of glycerol with butadiene to form glycerol ethers **1**, **2**, and **3**. *Reproduced from reference R. Palkovits, I. Nieddu, R.J.M. Klein Gebbink, B.M. Weckhuysen, Highly active catalysts for the telomerization of crude glycerol with 1,3-butadiene. ChemSusChem 1 (2008) 193, with kind permission.*

while Pd/TOMPP (TOMPP = *tris* (2-methoxyphenyl) phosphine) leads to the formation of mono-, di-, and triethers of glycerol, thus emphasizing the capability of Pd/TOMPP to telomerize sterically demanding nucleophiles as secondary alcohols [35]. Comparing the catalytic activity for the telomerisation of pure and crude glycerol, no significant difference in activity are observed.

3.5 POLYGLYCEROLS

Linear PGs obtained via catalytic oligomerization of glycerol, are biocompatible polyols of high thermal and chemical stability, which find growing utilization in cosmetic and food preparations as well as in technical applications [36].

The highly biocompatible nature of these oligomers has been fully assessed in the mid-2000s by in vitro as well as in vivo experiments and tests [37]. Results showed even higher biocompatibility when compared with some of the common biocompatible polymers already in human use. Due to their high thermal stability, PGs are ideally suited as plasticizers enabling higher polymer processing temperatures, e.g., in starch-based biodegradable thermoplastic compositions [38].

Regardless of their name, commercial PGs are rather comprised of mixtures going from (mainly) diglycerol (PG-2) up to (mainly) decaglycerol (PG-10), and should not be confused with truly polymeric PG formed by hyperbranched polymerization of glycidol. Such confusion may originate from using the word "polyglycerol" to indicate short-chain oligomers, first industrially manufactured starting from glycidol and, since the early 2000s, directly from glycerol when low-cost glycerol eventually became available.

Production makes use of alkali-catalyzed polycondensation of *n* glycerol molecules with elimination of *n*-1 water molecules (Scheme 3.5) [39].

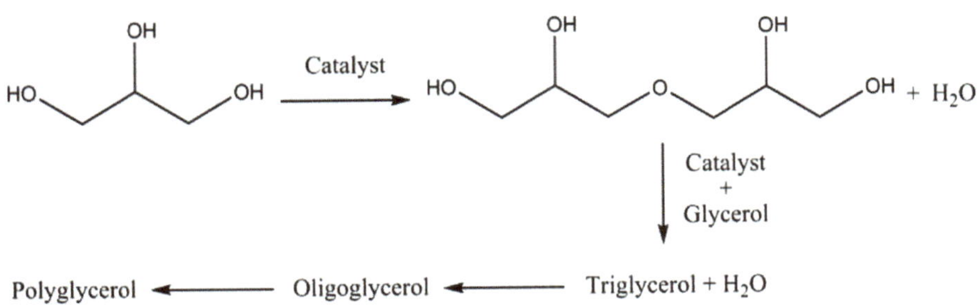

SCHEME 3.5

Schematic representation of the etherification of glycerol to linear polyglycerol (PG). *Reproduced from reference R.K. Kainthan, J. Janzen, E. Levin, D.V. Devine, D.E. Brooks, Biocompatibility testing of branched and linear polyglycidol, Biomacromolecules, 7, (2006) 703–709, with kind permission.*

The reaction is typically carried out at 260°C with 2.5 mol% catalyst (K_2CO_3, Li_2CO_3, KOH, NaOH) under nitrogen so as to exclude air and avoid acrolein formation, for 1–4 h, depending on the desired PG. The reaction temperature, catalyst nature and amount, and reaction time are the main parameters influencing the base-catalyzed polymerization [40]. The production cost of PG-3, PG-6, and PG-10 (3000/tonne in 1980 when glycerol was commercialized at $1490/tonne, and 60% of the production cost could be ascribed to glycerol) is independent of the degree of polymerization.

Homogeneous alkaline catalysis with carbonates is currently used in industry as it reduces the formation of cyclic oligomers normally used in place of hydroxides due to better solubility of carbonates in the glycerol and in the polymeric product [41]. Following removal of unreacted glycerol and water, the product is distilled.

Condensation may be intermolecular to produce linear oligomers, or intramolecular to give cyclic species. Lower reaction temperatures and lower pH favor the formation of cyclic isomers, whereas at higher temperatures side reactions produce dark colors and flavors of strong smell [42].

Today's technology allows to obtain diglycerol with less than 1% cyclic compound, or colorless PGs up to PG-6 with less than 5% cyclic compounds.

The reaction mechanism of the base-catalyzed polycondensation involves coordination of the primary hydroxyls of glycerol to the calcium ions with formation of a favored 6-membered ring, and easier attack of incoming deprotonated glycerol on the coordinated carbon of $-CH_2OH$ whose lengthened carbon oxygen bond lowers the energy of the transition state complex [43].

Aiming to improve commercial synthesis of PGs based on low amounts of CaO, Weckhuysen's team also developed a suitable method of catalyst immobilization [44]. The catalyst, comprised of CaO dispersed onto a carbon nanofiber support has higher activity than unsupported CaO, affording at 220°C a product with Gardner color number <2, containing no acrolein and minimal cyclic by-products (Fig. 3.9). The higher activity is due to the rapid formation and CaO colloids of smaller size compared with colloids from bulk CaO.

At 220°C the catalyst defragments and forms colloidal particles of pronounced etherification activity [45]. Colloidal CaO particles about 50–100 nm in size are spontaneously generated and their quantity gradually increases with increasing reaction time.

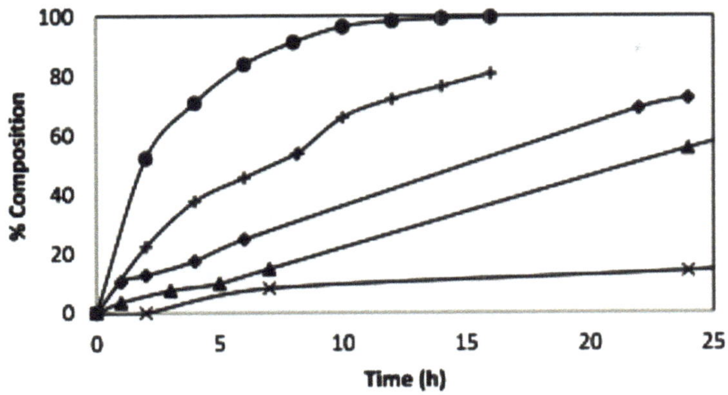

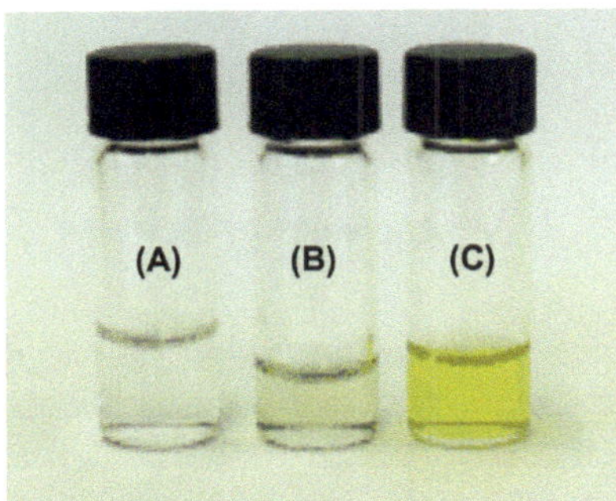

FIGURE 3.9
Glycerol conversion over 2 wt% CaO catalyst supported on carbon nanofiber, at 180°C (×), 200°C (▲), 220°C (♦), 240°C (+), 260°C (•). Picture shows product color at 70% conversion at (A) 220°C, (B) 240°C, and (C) 260°C. *Reproduced from reference F. Kirby, A.-E. Nieuwelink, B. W. M. Kuipers, A. Kaiser, P. C. A. Bruijnincx, B. M. Weckhuysen, CaO as Drop-In Colloidal Catalysts for the Synthesis of Higher Polyglycerols, Chem. Eur. J. 21 (2015) 5101–5109, with kind permission.*

Research in this field is flourishing. Recently, e.g., Jérôme's team in France in cooperation with Solvay (a primary manufacturer of PG) developed a catalytic oligomerization of glycerol route over a solid superacid catalyst (trade named *Aquivion PFSA*) affording targeted oligoglycerols with up to 75% yield [46].

The solid superacid catalyst, a sulfonic acid copolymer of tetrafluoroethylene and sulfonyl fluoride vinyl ether with a Hammett acidity function of −12, is not only highly selective and more active than commonly used solid acid catalysts, leading to oligoglycerols producing a reaction medium less colored than those obtained with traditional catalysts (Fig. 3.10), but also recyclable (successfully recycled 10 times without decrease of activity and selectivity) and more active than homogeneous catalysts.

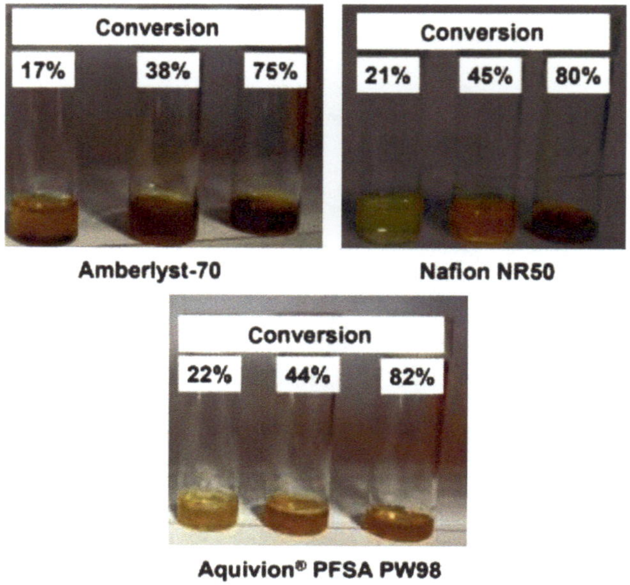

FIGURE 3.10
Color of the reaction mixture as a function of the conversion for Amberlyst-70, Nafion NR50, and Aquivion PFSA PW98. *Reproduced from reference A. Karama, N. Sayouda, K. De Oliveira Vigiera, J. Laib, A. Liebensb, C. Oldanic, et al., Heterogeneously-acid catalyzed oligomerization of glycerol over recyclable superacid Aquivion PFSA, J. Mol. Catal. A 422 (2016) 84–88, with kind permission.*

3.6 POLYGLYCEROL ESTERS

The hydroxyl groups of the PG molecule are easily esterified with carboxylic acids of different nature and in different PG:acid ratios to produce a wide variety of amphiphilic polyglycerol esters (PGEs) characterized by the desired HLB (hydrophilic–lipophilic balance) used across many industries due to their unique properties, such as high viscosity, retained also at high temperatures. PGEs indeed are increasingly used in a number of food products [47], in cosmetics and toiletries, where they outperform both homologous polyol esters and ethoxylated surfactants in reducing the interfacial tension.

For example, compared to those of commercial ethoxylates (*n*-dodecyl polyoxyethylene monoethers, $C_{12}EO_n$) the surfactant properties of polyglycerol monolaurates (PGML) in aqueous solution are distinctly better (Table 3.3): the foam is higher and more stable, and the reduction in interfacial tension, linked to better detergency, is higher [48]. Furthermore the surfactant properties of PGML having few glycerol units (di- to tetraglycerol monolaurates)

Table 3.3 Interfacial Tension and Detergency of Aqueous Solutions of Various Polyglycerol Monolaurate (PGML) and Ethoxylate Surfactants

Compound	Interfacial Tension (mN/m)	Detergency (%)
Blank	23.9	2
2GML	5.5	83.3
3GML	1.7	96.7
4GML	3.4	84.8
5GML	5.3	50
C-PGML	6.2	13.3
$C_{12}EO_4$	12.6	6.0
$C_{12}EO_6$	3.7	96
$C_{12}EO_8$	2.7	95.5

2GML, *diglycerol monolaurate;* 3GML, *triglycerol monolaurate;* 4GML, *tetraglycerol monolaurate;* 5GML, *pentaglycerol monolaurate;* C-PGML, *commercial polyglycerol monolaurate;* EO, *oxyethylene unit.*
Adapted from reference T. Kato, T. Nakamura, M. Yamashita, M. Kawaguchi, T. Kato, T. Itoh, Surfactant properties of purified polyglycerol monolaurates, J. Surfactants Deterg. 6 (2003) 331–337, with kind permission.

are similar to those of $C_{12}EO_n$ having many oxyethylene units (hexa- and octaoxyethylene).

Early applications of PGEs incorporated in polymers as antifogging, antistatic, and lubricating agents are now complemented by rapidly increasing use as safer and more ecological emulsifiers for making better and safer cleaning, personal care, and cosmetic products [49].

Driven by consumer demand to replace potentially toxic ingredients with naturally derived, safe alternatives, indeed, PGEs are ideally suited to stabilize oil-in-water (O/W) emulsions of natural oils such as vegetable-derived oils that, being composed of triglycerides, are notoriously difficult to emulsify requiring the use of undesirable ethoxylated surfactants.

Ethoxylated ingredients such as polyethylene glycol (PEG)-100 stearate are widely used as emulsifiers. PEG-based ingredients may contain toxic 1,4-dioxane as by-product of the industrial process used to make PEG [50].

The chemistry of PGEs offers unprecedented versatility to develop surface active agents that may vary considerably in composition owing to the possibility to tailor the extent of glycerol polymerization, the chemical nature of ester-forming acid, the degree of esterification and the possibility to use more than one type of hydrophobic group per molecule toward various application profiles. In addition, the physico-chemical properties of PGs are less sensitive to temperature and salts than those of PEGs.

Coupled to excellent biocompatibility, this has opened the route to their use as emulsifiers, dispersants, emollients, wetting agents, or thickeners in personal care formulations in cosmetic and personal care products, where they are increasingly replacing fatty acid and fatty alcohol ethoxylates not only by their natural origin, but also unique performance [51].

One example is Evonik's emulsifier ISOLAN GPS (International Nomenclature of Cosmetic Ingredients or INCI: Polyglyceryl-4 Diisostearate/Polyhydroxystearate/Sebacate) which provides good emulsion stabilization and additionally enables the formulation of fluid emulsions with very low oil phase ratios. This characteristic allows the realization of cost-efficient W/O emulsions with a pleasant and light skin feel [52].

Another, using higher molecular weight PGs, is the O/W emulsifier TEGO Care PSC 3 (INCI: Polyglyceryl-3 Dicitrate/Stearate), namely a mixed ester of PG with stearic acid and a substoichiometric, optimized amount of citric acid, suitable for formulations at a pH of 4.0–5.5 such as those using natural preservatives such as organic acids, in contrast to former products such as Glyceryl Stearate or Glyceryl Stearate Citrate [53]. Moreover, the structure and composition of Polyglyceryl-3 Stearate/Citrate were optimized to support the formation of liquid crystalline structures in emulsions to form O/W creams and lotions with an excellent stability profile without using polyacrylate-based thickeners.

In contrast, the O/W emulsifier tradenamed by Evonik TEGO Care PBS 6 (INCI: Polyglyceryl-6 Stearate and Polyglyceryl-6 Behenate) is a tailor-made solution for stabilization of low-viscous O/W emulsions even in combination with ingredients such as water-soluble UV filters [54]. The emulsifier indeed has an high hydrophilic–lipophilic balance (HLB = 13), which stabilizes O/W emulsions with high water content, namely low-viscous emulsions such as those used to formulate sunscreen, skin care, and body lotion as sprays, which are increasingly demanded by customers, showing at the same time excellent moisturization properties. Not only the new product of full vegetable origin replaces ethoxylated derivatives questioned for health and environmental reasons but it does so by providing enhanced performance.

3.7 GLYCEROL POLYESTERS

Synthesized via straightforward polycondensation reactions followed by a curing step, glycerol polyesters are biodegradable elastomers, which have been proposed as soft tissue replacement alternatives, due to the biocompatibility and biodegradable nature of constituent monomers. For example, poly

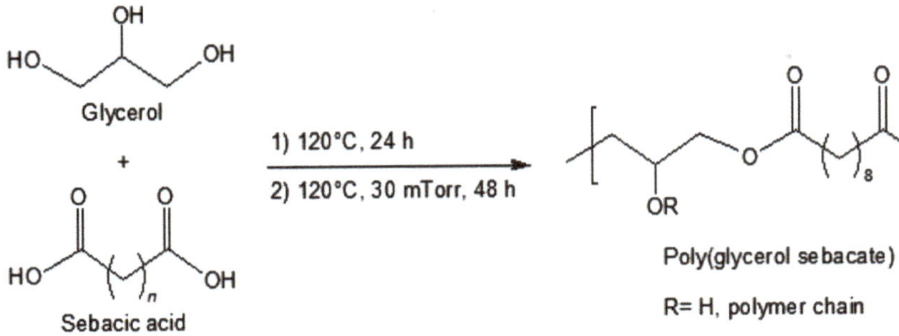

FIGURE 3.11
Reaction scheme for the synthesis of poly(glycerol sebacate) (PGS).

(glycerol sebacate) (PGS) synthesized from glycerol and sebacic acid (Fig. 3.11) [55], and then cured to the desired level of cross-linking enabling the tuning of physical properties [56], is a bioresorbable thermoset polyester with elastomeric properties.

Inherent elastomeric properties, the good biocompatibility (low degree of acute immune response upon implantation) [57] due to the compatibility with the Kreb's cycle of metabolism of its main components and degradation products (glycerol and sebacic acid), and the versatile esterification chemistry of glycerol (to tailor the degradation rate that is related to its hydrophobicity and flexibility) offer numerous possibilities for producing a variety of in vivo applications for cardiovascular, orthopedics, neurovascular, and tissue engineering. Dubbed *Regenerez*, indeed, the resin is produced and commercialized by an US-based company (Secant Medical) as new generation biomaterial [58].

Another breakthrough occurred in 2011 when it was discovered that glycerol and citric acid polymerize to form a thermoset resin, soluble in water, showing several important properties including quick degradation in the environment. Until the introduction of this thermoset, nearly all biodegradable plastics have been thermoplastic polymers. Dubbed "Plantics-GX" by the start-up company Plantics, the resin is currently produced on tonne scale at a pilot plant in the Netherlands.

In 2011 Gadi Rothenberg and Albert Alberts discovered by accident in their laboratory at the University of Amsterdam that combining citric acid dissolved in glycerol at a temperature above the boiling point of water (atmospheric pressure) and below 130°C gives a hard polyester resin by a straightforward Fisher esterification process. The resulting polymer is a "bio-bakelite," a hard polyester three-dimensional network. It adheres to almost everything, yet it is also hydro-degradable (by simple hydrolysis). Although in hindsight this is a very simple process, it has never been reported before and the material and its composites and applications were subsequently patented by the University of Amsterdam [59].

The extent of cross-linking is controlled by the reaction conditions, most notably temperature, reaction time, and glycerol:citric acid ratio. The higher the extent of cross-linking, the lower the rate of degradation in water. Highly cross-linked samples can survive for months in water, and indefinitely in air (the original sample from 2011 is still as good as new).

The boiling points of glycerol (290°C) and the decomposition temperature of citric acid (175°C) ensure that water is the only compound liberated as steam, as no decarboxylation takes place at $T < 150°C$.

When the reaction is complete (Fig. 3.12) the polymer does not decarboxylate even to high temperatures, though it slowly hydrolyzes in contact with water thereby requiring impermeabilization for use in industrial and technological applications.

The bio-polyester (Fig. 3.13) easily adheres to other materials and can therefore be used in combination with steel, glass, metals, and other solid materials used for making inflexible plastic items such as computer and telephone casings, insulation foam, trays, tables, and lamps. Composites made from Plantics-GX as matrix are exceptionally strong, because the polymer has a very large number of hydroxyl groups on its surface that can bind with practically any glass, metallic, or oxide fiber (though not with hydrophobic petro-based polymers such as polyethene or polypropene).

Alberts and Rothenberg also patented a "bio-Formica," namely new Plantics-GX polyester in combination with bio-based particulate or fibrous fillers to form composite materials (Fig. 3.14) [60].

The filler may be comprised of a cellulosic or lignocellulosic material such as wood chips, wood flakes, sawdust, paper pulp, or fibers such as cotton, linen, flax, and hemp.

Remarkably the resulting composite materials show a high fire resistance, which makes them suitable as building components and in other applications where fire-resistance is an issue. They are also inherently safe because they have no N atom and no S atoms, so there is no possibility of toxic gases.

The bio-polyester easily adheres to other materials and can therefore be used in combination with steel, glass, metals, and other solid materials used for making inflexible plastic items such as computer and telephone casings, insulation foam, trays, tables, and lamps.

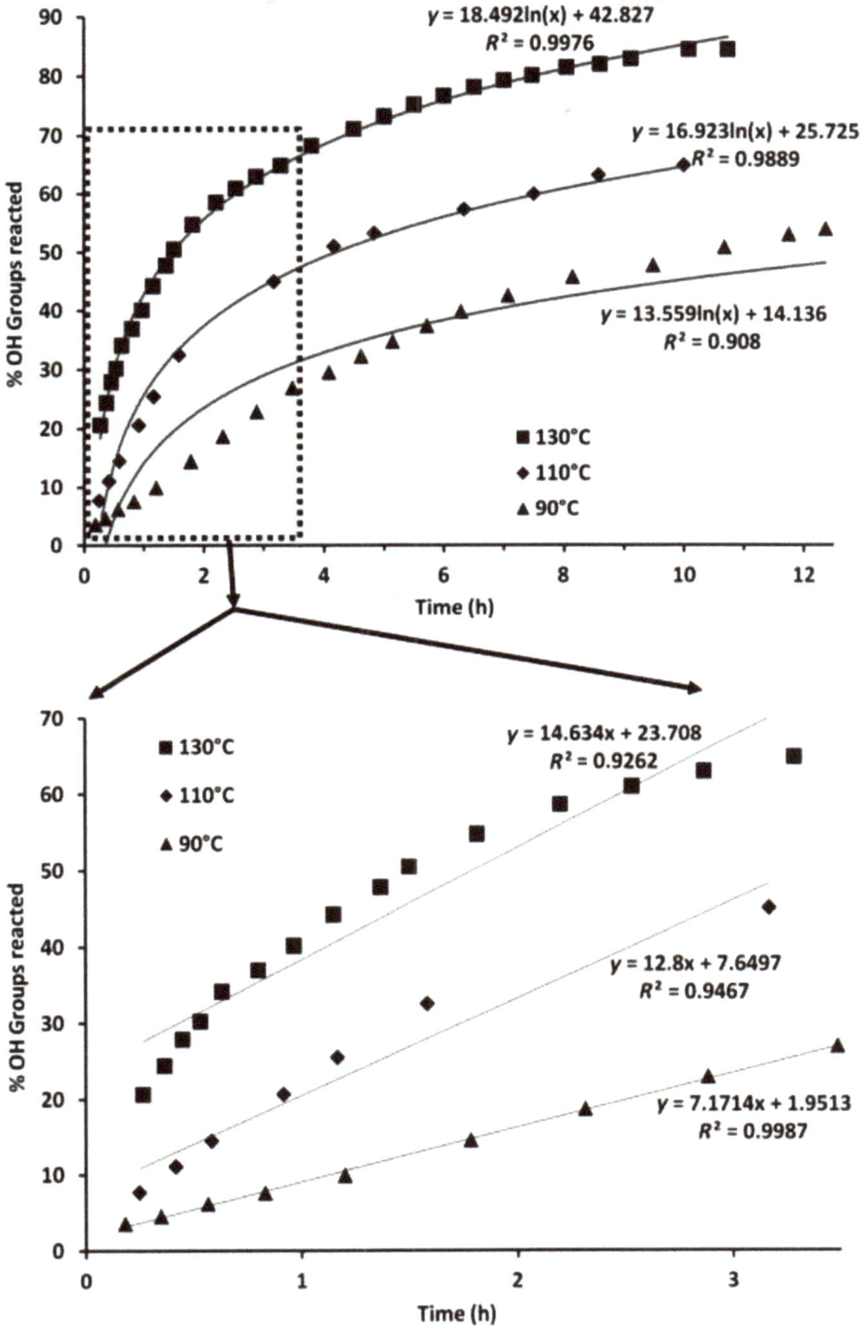

FIGURE 3.12

Kinetics of the reaction of glycerol and citric acid in 1:1 molar ratio at 130°C, 110°C, and 90°C. Top: A logarithmic relationship is observed for esterification reactions up to 12.5 h. Polymers fabricated at 130°C show the most accurate logarithmic profile. Bottom: Linear relationships are observed for esterification reactions for the first 3.5 h. Polymers fabricated at 90°C have the most linear control in the first 3.5 h. *Reproduced from reference J.M. Halpern, R. Urbanski, A.K. Weinstock, D.F. Iwig, R.T. Mathers, H.A. von Recum, A biodegradable thermoset polymer made by esterification of citric acid and glycerol, J. Biomed. Mater. Res. A. 102 (2014) 1467–1477, with kind permission.*

FIGURE 3.13

A sample of Plantics-GX, the thermoset resin obtained by glycerol and citric acid polycondensation. *Pictures of Professor Gadi Rothenberg, reproduced with kind permission.*

FIGURE 3.14

A sample of Plantics-GX-based polyester in combination with wood to form composite material. *Pictures of Professor Gadi Rothenberg, reproduced with kind permission.*

Remarkably the resulting composite materials show a high fire resistance, which makes them suitable as building components and in other applications where fire-resistance is an issue. Full biodegradability ensures that the composite can be disposed of as organic waste as the material hydrolyzes in water making the bio-based particulate available for biological degradation.

The new polymer may function as alternative to replace PUs, polystyrene, and epoxy resins. It can also be used for advanced applications where the good health and safety profile is required, such as in the case of drug delivery applications, as shown by good antibiotic activity of polymer-entrapped gentamicin against *Staphylococcus aureus* bacteria [61].

It can also be used as environmental friendly and high-performance coating for improving the technical performance of outdoor wood products used in construction wherein a citric acid–glycerol mixture (CA-G) increases hardness and decay resistance, including resistance to swelling (up to 53%) [62].

The company (Plantics) is currently working with several partners on market applications, and multiton production is foreseen for 2018–20.

Industrial production will start following customer demand of large quantities. Right now Plantics is making tonnes of the material. Most likely Plantics will make up to a few hundred tonnes, but large-scale production will be probably done with partners at the respective location. Because Plantics-GX is made from simple starting materials that are available practically everywhere, it makes sense to make it on location.

As to the size of the plastics market that Glycix can reasonably impact in 5 years from now, perhaps 1%–5% of the thermoset resin market could be hit. "The plastics industry is very conservative," says Rothenberg, "so it will take time before companies make the move from plastics to Plantics."

Other bio-based carboxylic acids can be used to produce polyesters with glycerol affording polymers with remarkable properties. One relevant example is succinic acid affording PG succinate, a polyester showing rubbery behavior even with glycerol of minimum 95% purity (i.e., not only with pure glycerol) [63].

3.8 POLYGLYCEROLS: LARGE GROWTH POTENTIAL

Referring to the rapidly increasing number of publications on fatty acid PGEs between the early 1980s and the early 2000s, in 2007 Pérez Pariente and coworkers were noting how these materials continued to remain "virtually unexplored outside the industrial R&D departments." A decade later the abundant availability of glycerol at low cost coupled to the excellent performance of both PGs and PGEs to replace a number of oil-derived surface active agents has led to dramatic increase of market size, and research activities.

Segmented into personal care, food and beverage, packaging and pharmaceutical industry, the $2.4 billion PG market in 2014 is expected to grow at 5.5% rate from 2015 to 2020 due to rising demand in all sectors, including surfactants, personal care products ecological polymer additives, antifogging films, and pigment dispersants [64].

Original manufacturers of PGs including Sakamoto in Japan and Solvay Chemicals and Lonza in Europe, saw the entrance of a number of new companies including Lonza, Brenntag, Danisco, Estelle, and many others. In India companies such as Fine Organics, Parry Enterprises, and Estelle Chemicals polymerize glycerol and manufacture large amounts of PGE of fatty acids as emulsifiers and surfactants for value-added applications in food and cosmetics.

In Europe (noncomprehensive list), Spiga NORD, Lonza, Solvay, Akzo Nobel, Evonik, Clariant, and BASF manufacture PGs and PGEs. In China a recent market research report [65] on polyglycerol-3 industry alone, included 10 different suppliers of PG, led by Hangzhou J&H Chemical, while other companies such as Silver-Un Chemical Technology manufacture other PGs, such as polyglycerol-10.

Consumers demand functional products containing nontoxic ingredients, preferably of natural origin; even though at times the ingredient to be replaced from industrial food will be polyglycerol polyricinoleate (PGPR; formed via esterification of castor oil fatty acids with PG) as a large chocolate company (Hershey) is currently doing in the United States [66].

PGPR is a powerful water-in-oil emulsifier used by the chocolate industry as an emulsifier replacing cocoa butter, which is extracted and sold to the cosmetic industry at higher price, even though cocoa butter contains most of the highly beneficial polyphenols of chocolate.

The global food emulsifier market is a rapidly growing segment within the food ingredients market due to the growing trend toward reducing fat content in food products. This market is largely driven by diglycerides, lecithin, stearoyl lactylates, and other emulsifiers such as PGEs, PGPR, polysorbate, and sucrose esters.

Now emulsifiers added to food increase intestinal permeability by breaching the integrity the intestinal epithelial barrier that, with its intercellular tight junction, controls the equilibrium between tolerance and immunity to non-self antigens. This effect might explain the rising incidence of autoimmune disease during the last decades [67].

References

[1] Frost & Sullivan, World Fatty Ester Markets, 2004.

[2] Grand View Research, Fatty Acid Ester Market, San Francisco, 2016.

[3] EU directive 2006/52/EC of 5 July 2006 amending Directive 95/2/EC on food additives other than colours and sweeteners and Directive 94/35/EC on sweeteners for use in foodstuffs.

[4] K. Kaur, R.K. Wanchoo, A.P. Toor, Sulfated iron oxide: a proficient catalyst for esterification of butanoic acid with glycerol, Ind. Eng. Chem. Res. 54 (2015) 3285–3292.

[5] H. Bjornberg, R.C. Jansson, A waterborne binder composition and use thereof, EP 2064295 A1, (2009).

[6] K.V. Driessche, K. Schwarzer, R. Sygall, F. Haesebrouck, R. Ducatelle, F.V. Immerseel, Monobutyrin and trivalerin as feed additives reduce *Salmonella enteritidis* colonization in chickens, 3rd IHSIG International Symposium on Poultry Gut Health, Ghent, Belgium, 2015.

[7] D.R. Satriana, N. Arpi, Y.M. Lubis, Adisalamun, M.D. Supardan, W.A.W. Mustapha, Diacylglycerol-enriched oil production using chemical glycerolysis, Eur. J. Lipid Sci. Technol. 118 (2016) 1880–1890, <http://dx.doi.org/10.1002/ejlt.201500489>.

[8] H. Takano, Y. Itabashi, Molecular species analysis of 1,3diacylglycerols in edible oils by HPLC/ESI-MS, Bunseki Kagaku 51 (2002) 437–442.

[9] M. Matsuo, Nutritional characteristics and health benefits of diacylglycerol in foods, J. Food Sci. Technol. 10 (2004) 103–110.

[10] L. Vázquez, N. González, G. Reglero, C. Torres, Solvent-free lipase-catalyzed synthesis of diacylgycerols as low-calorie food ingredients, Front. Bioeng. Biotechnol 4 (2016) 6.

[11] M. Damstrup, A.D. Jensen, S. Kiil, F.V. Sparsø, X. Xu, Continuous glycerolysis in an immobilized enzyme packed reactor for industrial monoacylglycerol production, 99th AOCS Annual Meeting & Expo, Seattle, WA, May 18, 2008.

[12] J. Aracil, M. Vicente, M. Martinez, M. Poulina, Biocatalytic processes for the production of fatty acid esters, J. Biotechnol. 124 (2006) 213–223.

[13] H. Taguchi, H. Watanabe, K. Onizawa, T. Nagao, N. Gotoh, T. Yasukawa, et al., Double-blind controlled study on the effects of dietary diacylglycerol on postprandial serum and chylomicron triacylglycerol responses in healthy humans, J. Am. Coll. Nutr. 19 (2000) 789–796.

[14] EFSA, Opinion of the scientific panel on dietetic products, nutrition and allergies on a request from the commission related to an application to market enova oil as a novel food in the EU, EFSA J. 159 (2004) 1–19.

[15] Bundeninstitut fur Risikobewertung. Infant formula and follow-up formula may contain harmful 3-MCPD fatty acid esters. BfR Opinion No. 047/2007. Available online: <http://www.bfr.bund.de/cm/349/infant_formula_and_follow_up_formula_may_contain_harmful_3_mcpd_fatty_acid_esters.pdf>, 2007.

[16] Kao Corporation, Kao to Temporarily Refrain from Selling *Econa* Products, kao.com, September 16, 2009.

[17] AOCS, Kao Halts Sales of DAG Oil, September 16, 2009.

[18] Agriculture and Agri-Food Canada, Functional foods and beverages in Japan, Global Analysis Report, Ottawa: April 2016.

[19] C. Li, X. Luo, T. Li, X. Tong, Y. Li, Polyurethane foams based on crude glycerol-derived biopolyols: one-pot preparation of biopolyols with branched fatty acid ester chains and its effects on foam formation and properties, Polymer 55 (2014) 6529–6538.

[20] M.O. Sonnati, S. Amigoni, E.P. Taffin de Givenchy, T. Darmanin, O. Choulet, F. Guittard, Glycerol carbonate as a versatile building block for tomorrow: synthesis, reactivity, properties and applications, Green Chem. 15 (2013) 283–306.

[21] E. Schols, Production of cyclic carbonates from CO_2 using renewable feedstocks, CEOPS Workshop "R&D on CO2 Utilization", May 26, 2014, Lille, France.

[22] J.R. Ochoa-Gómez, O. Gómez-Jiménez-Aberasturi, C. Ramírez-López, M. Belsué, A. Brief, Review on industrial alternatives for the manufacturing of glycerol carbonate, a green chemical, Org. Process Res. Dev. 16 (2012) 389–399.

[23] S. Claude, Z. Mouloungui, J.-W. Yoo, A. Gaset, Method for preparing glycerol carbonate, US6025504, 1999.

[24] L. Wang, Y. Ma, Y. Wang, S. Liu, Y. Deng, Efficient synthesis of glycerol carbonate from glycerol and urea with lanthanum oxide as a solid base catalyst, Catal. Commun. 12 (2011) 1459–1462.

[25] A. Galadima, O. Muraza, Sustainable Production of Glycerol Carbonate from By-product in Biodiesel Plant, Waste Biomass Valor 88 (2017) 141–152.

[26] N. Nguyen, Y. Demirel, Economic analysis of biodiesel and glycerol carbonate production plant by glycerolysis, J. Sustain. Bioen. Syst 3 (2013) 209–216.

[27] N.A.S. Zuhaimi, V.P. Indran, M.A. Deraman, N.F. Mudrikah, G.P. Maniam, Y.H. Taufiq-Yap, et al., Reusable gypsum based catalyst for synthesis of glycerol carbonate from glycerol and urea, Appl. Catal. A 502 (2015) 312–319.

[28] M.J. Climent, A. Corma, P.D. Frutos, S. Iborra, M. Noy, A. Velty, et al., Chemicals from biomass: synthesis of glycerol carbonate by transesterification and carbonylation with urea with hydrotalcite catalysts. The role of acid–base pairs, J. Catal. 269 (2010) 140–149.

[29] S. Holmiere, R. Valentin, P. Marechal, Z. Mouloungui, Esters of oligo-(glycerol carbonate-glycerol): new biobased oligomeric surfactants, J. Coll. Interf. Sci 487 (2017) 418–425.

[30] M. Sutter, E. Da Silva, N. Duguet, Y. Raoul, E. Métay, M. Lemaire, Glycerol ether synthesis: a bench test for green chemistry concepts and technologies, Chem. Rev. 115 (2015) 8609–8651.

[31] A. Behr, L. Obendorf, Development of a process for the acid-catalyzed etherification of glycerine and isobutene forming glycerine tertiary butyl ethers, Eng. Life Sci. 2 (2003) 185–189.

[32] J.A. Melero, G. Vicente, G. Morales, M. Paniagua, J.M. Moreno, R. Roldàn, et al., Acid-catalyzed etherification of bio-glycerol and isobutylene over sulfonic mesostructured silicas, Appl. Catal., A 346 (2008) 44.

[33] Z. Fan, Y. Zhao, F. Preda, J.-M. Clacens, H. Shi, L. Wang, et al., Preparation of bio-based surfactants from glycerol and dodecanol by direct etherification, Green Chem. 17 (2015) 882–892.

[34] R. Palkovits, I. Nieddu, R.J.M. Klein Gebbink, B.M. Weckhuysen, Highly active catalysts for the telomerization of crude glycerol with 1,3-butadiene, ChemSusChem 1 (2008) 193.

[35] R. Palkovits, I. Nieddu, C.A. Kruithof, R.J.M. Klein Gebbink, B.M. Weckhuysen, Palladium-based telomerization of 1,3-butadiene with glycerol using methoxy-functionalized triphenylphosphine ligands, Chem. Eur. J. 14 (2008) 8995.

[36] R. Ciriminna, B. Katryniok, S. Paul, F. Dumeignil, M. Pagliaro, Glycerol-derived renewable polyglycerols: a class of versatile chemicals of wide potential application, Org. Process Res. Dev. 19 (2015) 748–754.

[37] R.K. Kainthan, J. Janzen, E. Levin, D.V. Devine, D.E. Brooks, Biocompatibility testing of branched and linear polyglycidol, Biomacromolecules 7 (2006) 703–709.

[38] A. Taghizadeh, P. Sarazin, B.D. Favis, High molecular weight plasticizers in thermoplastic starch/polyethylene blends, J. Mater. Sci. 48 (2013) 1799–1811.

[39] A. Martin, M. Richter, Oligomerization of glycerol—a critical review, Eur. J. Lipid Sci. Technol. 113 (2011) 100–117.

[40] N. Garti, A. Aserin, B. Zaidman, Polyglycerol esters: optimization and technoeconomic evaluation, J. Am. Oil Chem. Soc 58 (1981) 878–883.

[41] C. Marquez-Alvarez, E. Sastre, J. Pérez Pariente, Solid catalysts for the synthesis of fatty esters of glycerol, polyglycerols and sorbitol from renewable resources, Top. Catal 27 (2004) 105–117.

[42] G.L. Hasenhuettl, Synthesis and commercial preparation of food emulsifiers, in: G.L. Hasenhuettl, R.W. Hartel (Eds.), Food Emulsifiers and Their Applications, Springer Science + Business Media, New York, 2008, pp. 11–37.

[43] S. Salehpour, M.A. Dube, Towards the sustainable production of higher-molecular-weight polyglycerol, Macromol. Chem. Phys. 212 (2011) 1284–1293.

[44] F. Kirby, A.-E. Nieuwelink, B.W.M. Kuipers, A. Kaiser, P.C.A. Bruijnincx, B.M. Weckhuysen, CaO as drop-in colloidal catalysts for the synthesis of higher polyglycerols, Chem. Eur. J. 21 (2015) 5101–5109.

[45] A.M. Ruppert, J.D. Meeldijk, B.W. Kuipers, B.H. Erné, B.M. Weckhuysen, Glycerol etherification over highly active CaO-based materials: new mechanistic aspects and related colloidal particle formation, Chem. Eur. J 14 (2008) 2016.

[46] A. Karama, N. Sayouda, K. De Oliveira Vigiera, J. Laib, A. Liebensb, C. Oldanic, et al., Heterogeneously-acid catalyzed oligomerization of glycerol over recyclable superacid Aquivion PFSA, J. Mol. Catal. A 422 (2016) 84–88.

[47] N.J. Krog, Food emulsifiers and their chemical and physical properties, in: S.E. Friberg, K. Larsson (Eds.), Food Emulsions, Marcel Dekker, New York, 1997, pp. 141–188.

[48] T. Kato, T. Nakamura, M. Yamashita, M. Kawaguchi, T. Kato, T. Itoh, Surfactant properties of purified polyglycerol monolaurates, J. Surfactants Deterg. 6 (2003) 331–337.

[49] H.H. Wenk, J. Meyer, Polyglycerol—a versatile building block for sustainable cosmetic raw materials, SOFW J. 135 (8) (2009) 25–30.

[50] M.L.W. Juhász, E.S. Marmur, A review of selected chemical additives in cosmetic products, Dermatol. Ther. 27 (2014) 317–322.

[51] H. Szeląg, E. Sadecka, R. Pawłowicz, A. Kuziemska, Emulsifiers from renewable materials: an eco-friendly synthesis and properties, Pol. J. Chem. Tech 15 (2013) 128–135.

[52] J. Meyer, P. Allef, H. Foetsch, A novel PEG-free emulsifier designed for formulating W/O lotions with a light skin feel, SOFW J. 131 (11) (2005) 20.

[53] B. Jha, J. Meyer, G. Polak, Integrating natural, sustainable and performance characteristics in personal care products, SOFW J. 136 (9) (2010) 36.

[54] A. Friedrich, P. Biehl, F. Unger, J. Marian von Hof, J. Meyer, Personal Care, March 2014, pp. 33–37.

[55] Y. Wang, G.A. Ameer, B.J. Sheppard, R. Langer, A tough biodegradable polymer, Nat. Biotechnol. 20 (2002) 602–606.

[56] R. Rai, M. Tallawi, A. Grigore, A. Boccaccini, Synthesis, properties and biomedical applications of poly(glycerol sebacate) (PGS): a review, Progr. Polym. Sci. 37 (2012) 1051–1078.

[57] I. Pomerantseva, N. Krebs, A. Hart, C.M. Neville, A.Y. Huang, C.A. Sundback, Degradation behavior of poly(glycerol sebacate), J. Biomed. Mater. Res. A 91 (2009) 1038–1047.

[58] I. Pomerantseva, N. Krebs, A. Hart, C.M. Neville, A.Y. Huang, C.A. Sundback, Degradation behavior of poly(glycerol sebacate), J. Biomed. Mater. Res. A 91 (2009) 1038–1047.

[59] A.H. Alberts, G. Rothenberg, Process for Preparing Foamed Polymer, WO 2012052385, (2012).

[60] A.H. Alberts, G. Rothenberg, Composite Material Comprising Synthetic Filler and Specific Polymer, WO2012/140239, (2012).

[61] J.M. Halpern, R. Urbanski, A.K. Weinstock, D.F. Iwig, R.T. Mathers, H.A. von Recum, A biodegradable thermoset polymer made by esterification of citric acid and glycerol, J. Biomed. Mater. Res. A 102 (2014) 1467–1477.

[62] G.G. Essoua Essoua, P. Blanchet, V. Landry, R. Beauregard, Pine wood treated with a citric acid and glycerol mixture: biomaterial performance improved by a bio-byproduct, Bioresources 11 (2016) 3049–3072.

[63] O. Valerio, T. Horvath, C. Pond, M. Misra, A. Mohanty, Improved utilization of crude glycerol from biodiesel industries: synthesis and characterization of sustainable biobased polyesters, Ind. Crops Prod. 78 (2015) 141–147.

[64] Grand View Research, polyglycerol market analysis, market size, application analysis, regional outlook, competitive strategies, and forecasts, 2015 to 2022, San Francisco, 2015.

[65] Prof Research, Global and Chinese Polyglycerin-3 (CAS 25618-55-7) Industry, 2009–2019 Market Research Report, Shanghai, 2014.

[66] J. Scipioni, Hershey's remake of 'The Great American Chocolate Bar,' Fox Business, June 17, 2015.

[67] A. Lerner, T. Matthias, Changes in intestinal tight junction permeability associated with industrial food additives explain the rising incidence of autoimmune disease, Autoimmun. Rev. 14 (2015) 479–489.

CHAPTER 4

Antifreeze and Multipurpose Cement Aid

4.1 GLYCEROL AS ANTIFREEZE

Pure glycerol, with a freezing point of approximately 18°C, is rarely seen in crystalline state because even the small amounts of water usually present in most formulations greatly depress its freezing point. A combination of two parts of glycerin to one part of water forms a eutectic mixture which freezes at −46.5°C. The complete concentration–temperature state diagram of water/glycerol mixtures with its typical V-shaped curve (Fig. 4.1) was first published by Lane, a chemist working at a soap making company, in 1925, emphasizing how [1]:

> It is of interest to note that a 56 per cent by weight glycerol solution, a concentration commonly used as an antifreeze in automobile radiators, was sealed in a glass capsule and taken to −72°C without bursting the tube. This indicates that there is little or no expansion with this concentration, even when completely frozen.

Indeed, due to its excellent antifreeze properties and nontoxic nature, glycerol was the first permanent-type antifreeze for automobile radiator cooling systems as early as 1920, to be eventually replaced by ethylene glycol in the 1930s due to lower cost of mono-ethylene glycol (MEG) and to the request of glycerol for making nitroglycerin in war times [2].

Adding glycerol to water stabilizes the liquid state because hydrogen bonding between water and glycerol increases the nucleation barrier for ice crystallization, decreasing the water hydrogen bonding ability, resulting in a similar effect to applying pressure to it [3].

An internal combustion engine generates a significant amount of heat, which is dissipated in the radiator using a coolant, thereby preventing engines from overheating.

Antifreeze coolants are used to ensure the transfer of heat under extreme cold conditions. We remind here that water, with its large heat capacity, is an excellent heat transfer fluid, but since it freezes at 0°C, aqueous formulations

Glycerol. DOI: http://dx.doi.org/10.1016/B978-0-12-812205-1.00004-7

Freezing Points of Glycerol and Its Aqueous Solutions[1]

By Leonard B. Lane

Armour Soap Works, Chicago, Ill.

THE use of glycerol as an antifreeze solution called for more information as to the freezing points of various mixtures of glycerol and water than was available in the literature. The Bureau of Standards[2] has published data showing the freezing points of solutions with concentrations as high as 50 per cent by volume of glycerol, the specific gravity of which was 1.2537 at 15° C., equivalent to 95.42 per cent glycerol by weight at 15°/15° C. according to Gerlach's table. The investigation was evidently not carried beyond this point. As the curve at this lowest temperature indicated that more concentrated solutions had lower freezing points, a study was started to determine the curve for all concentrations. The values recorded as the freezing points were the points at which crystals first appeared and continued to grow. As glycerol has a great tendency to supercool, it was found that seeding was required to obtain the true freezing points of all the more concentrated solutions beginning at the eutectic point.

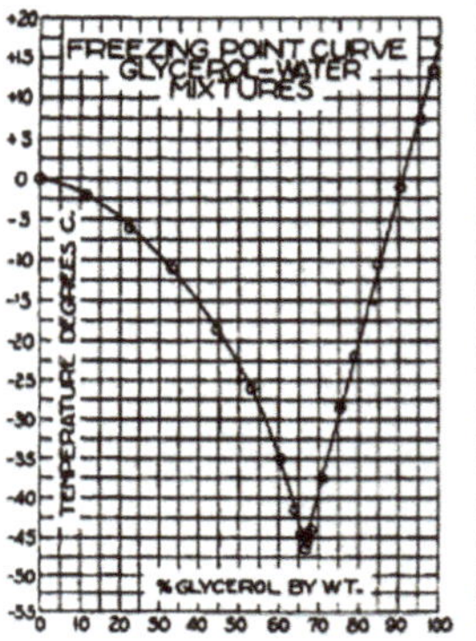

The method employed to reach low temperatures was that of refrigeration with carbon dioxide snow. The glycerol was immersed in a bath of alcohol, which was suspended in a larger container, and into this was sprayed liquid carbon dioxide from an inverted carbon dioxide cylinder. The ensemble was well insulated by felt and an air jacket.

It is of interest to note that a 56 per cent by weight glycerol solution, a concentration commonly used as an antifreeze in automobile radiators, was sealed in a glass capsule and taken to −72° C. without bursting the tube. This indicates that there is little or no expansion with this concentration, even when completely frozen.

Table I—Freezing Points of Glycerol-Water Mixtures

Glycerol by weight Per cent	Water Per cent	Freezing points ° C.	Glycerol by weight Per cent	Water Per cent	Freezing points ° C.
0.0[a]	100.0	0	65.0	35.0	−43.0
5.0	95.0	− 0.6	65.6[b]	34.4	−44.5
10.0	90.0	− 1.6	66.0[b]	34.0	−44.7
11.5[b]	88.5	− 2.0	66.7[b]	33.3	−46.5
15.0	85.0	− 3.1	67.1[b]	32.9	−45.5
20.0	80.0	− 4.8	67.3[b]	32.7	−44.5
22.6[b]	77.4	− 6.0	68.0[b]	32.0	−44.0
25.0	75.0	− 7.0	70.0	30.0	−38.9
30.0	70.0	− 9.5	70.9[b]	29.1	−37.5
33.3[b]	67.0	−11.0	75.0	25.0	−29.8
35.0	65.0	−12.2	75.4[b]	24.6	−28.5
40.0	60.0	−15.4	79.0[b]	21.0	−22.0
44.3[b]	55.5	−18.5	80.	20.0	−20.3
45.0	55.0	−18.8	84.8[b]	15.2	−10.5
50.0	50.0	−23.0	85.0	15.0	−10.9
53.0[b]	47.0	−26.0	90.0	10.0	− 1.6
55.0	45.0	−28.2	90.3[b]	9.7	− 1.0
60.0	40.0	−34.7	95.0	5.0	7.7
60.4[b]	39.6	−35.0	95.3[b]	4.7	7.3
64.0[b]	36.0	−41.5	98.2[b]	1.8	13.5
64.7[b]	35.3	−42.5	100.0[a]	0.0	17.0

[a] Taken from literature.
[b] Actual determination.
Remaining values were interpolated from curve.

Table II—Bureau of Standards Values Compared with Values Interpolated from Curve

Per cent glycerol by volume of 95.42 per cent by weight	Per cent glycerol by weight	Freezing Points: B. of S. ° C.	Freezing Points: From curve ° C.
10	11.7	− 2	− 2.0
20	22.8	− 6	− 5.8
30	33.4	−11·	−11.5
40	43.4	−18	−17.7
50	53.0	−26	−26.0

[1] Received June 15, 1925.
[2] *Letter Circular 88.*

FIGURE 4.1

The frontispiece of Lane's article with the V-shaped state diagram of water/glycerol mixtures. *Reproduced from reference L.B. Lane, Freezing points of glycerol and its aqueous solutions, Ind. Eng. Chem. 17 (1925) 924, with kind permission.*

of MEG and mono-propylene glycol (MPG) are commercially used as secondary working fluids to protect system piping and components in refrigeration, ventilation, and air conditioning systems from damage caused by freezing (volume expansion due to ice formation can easily rupture pipes and destroy equipment) [4].

Glycerol as heat transfer fluid formulated with water and antimicrobial agents meets demanding performance requirements in terms of effective freeze protection (freezing point below −40°C and efficient heat transfer over broad temperature range).

Glycerol-based antifreeze formulations present no harmful effects on health and environment, being generally noncorrosive to metal pipe materials and being available at low cost. Furthermore they are less flammable than MPG-based formulations. Since 2014, e.g., the Y-12 National Security Complex enriching uranium in Tennessee, has replaced the 50% solution of MPG with

48% solution of glycerol for all its 48 antifreeze loop systems, even though care was needed to replace the former solution with the much more hygroscopic glycerin solution [5].

4.2 GLYCEROL-BASED ANTIFREEZE MARKET

Glycerol is highly suitable as antifreeze ingredient, this being one of the functions of "man's most versatile chemical servant" identified by Milton Lesser in 1949 [6]. Antifreeze applications include engine coolant, plumbing fluid and heat transfer fluids.

Yet when the price of glycerol suddenly went down the price of MEG in 2006 immediately the producers of conventional antifreeze reacted claiming that "glycerine won't replace MEG as it simply doesn't have the technical properties" [7]. Research, however, readily restarted and the excellent antifreeze properties of glycerol/water mixtures were confirmed also for today's internal combustion engines [8].

In early 2010 "in order to remove technical barriers that currently prevent the cost-effective replacement of ethylene glycol with more environmentally friendly glycerine in antifreeze" the American Society for Testing and Materials (ASTM) announced the forthcoming introduction of engine coolant standards on natural glycerol in the United States [9].

A Glycerine Task Force Group was formed, comprising participants from the oleochemical/glycerin and engine coolant industries, which worked to determine the appropriate grade of refined glycerin required to ensure compatibility with engine parts and seals, and then what impurities were present in glycerin to be controlled and measured.

The project reached approval stages in early 2011 and in May the first ASTM specification for engine coolant grade glycerol (D7640) identifying the chemical and physical property requirements of the engine coolant grade glycerol, including its 99.5% purity, was published along with the D7714 and D7715 specifications for light and heavy-duty vehicle coolant grades.

In 2016 all standards were reapproved as ASTM D7640 - 16 "Standard Specification for Engine Coolant Grade Glycerine"; ASTM D7714 - 11(2016)e1 "Standard Specification for Glycerine Base Engine Coolant for Automobile and Light-Duty Service"; and ASTM - D7715 - 12(2016)e1 "Standard Specification for Fully-Formulated Glycerine Base Engine Coolant for Heavy-Duty Engines."

Two more new standards, ASTM D7637 - 10(2015), "Test Method for Determination of Glycerine Assay by Titrimetric (Sodium Meta Periodate)"; and ASTM D7638 - 10(2015) "Test Method for Determination of Fatty Acids and Esters in Glycerine" specifying the analytical methods for certification had been reapproved in 2015.

One such product meeting the ASTM D7640 standard is VYCERIN GL89 commercialized by Vantage Oleochemicals. On commercializing it, the company certifies the following: Methanol ≤ 0.1%; Ash ≤ 0.01%; Iron ≤ 2 ppm; Chloride ion ≤ 5 ppm; Sulfate ion ≤ 10 ppm; Silicon ≤ 10 ppm; Boron ≤ 10 ppm; Total Nitrite, Nitrate, and Phosphate ≤ 10 ppm; Total Aluminum, Calcium, Copper, Iron, Magnesium, Lead, Zinc ≤ 5 ppm.

Not surprisingly a number of antifreeze products based on "green" and "natural" glycerol were readily marketed in the United States and western Europe first, and in Canada and in Russia later.

A market report in late 2014 projecting the global automotive coolant market to reach a value of $10.9 billion by 2019 comprised glycerin along with MEG and MPG as one of the only three product types [10]. Still in 2014 more than 25% of MEG produced in the world was used in coolant mixtures [11].

Another business report convened that glycerol was the only alternative to MPG and MEG, noting that the market, while comprised of several major well-established companies, requires newer entrants to adopt innovation and differentiation strategies to differentiate themselves from the competition, with nontoxicity, high compatibility, and low cost having been the driving forces that led MPG to become the leading ingredient surpassing MEG in sales in recent years [12].

4.3 RAW BIOGLYCEROL AS MULTIPURPOSE CEMENT AID

Following a patent granted in 2006 [13], in 2008 Rossi's and Pagliaro's teams in Italy reported in the scientific literature the discovery that crude biodiesel glycerol is a multipurpose admixture ingredient [14]. The addition of crude bioglycerol to clinker, indeed, considerably enhances the strength of hardened concrete, by improving its resistance to compression, eases the clinker grinding process, and affords cement with better resistance to cracking, and water penetration.

Usually present as blends in finished cement particles, or topically applied products after cement application, admixtures are used in more than 60% of world cement production to save energy, reduce cracking, corrosion as well as to speed up the construction process for saving money and energy.

Admixtures are generally comprised of water soluble organic chemicals (typically, glycols and alkanolamines) and are added in relatively small quantity to improve the quality and manageability of the fresh and hardened concrete, and for achieving other desired properties [15]. The common chemical admixtures are classified as set retarding, air entrainment, water reducing, set accelerating, shrinkage reducing, superplasticizers, and corrosion inhibiting admixture [16].

In general the benefits of using bioglycerol instead of oil-derived cement additives conventionally used by industry, are significant for both environment and industry. The advantages for industry derive from having a single, readily available material that offers all three major technical improvements required of cement additives, namely (1) enhanced concrete strength; (2) grinding aid; and (3) handling aid.

In addition, beyond enabling the access at lower cost to higher grades cements, bioglycerol also provides important positive effects on the final cement imparting beneficial properties such as corrosion resistance and waterproofing.

As explained in a 2015 interview [17]:

> Bioglycerol is a most valuable compound for the concrete construction industry, exactly for the same reason which makes it unusable in pharmaceutical or personal care applications, namely that it is not pure.
>
> Residues of soap as well as of α-tocopherol from the biodiesel production, lead to excellent mechanical, anticorrosive properties and to better performance as a cement grinding aid than both pure glycerol and oil-derived additives. Reduction in power consumption in the grinding process is up to 10% when compared to oil-derived additives, while concrete finished structures are more than 10% stronger than identical structures obtained with no added bioglycerol.
>
> Finally, the stability of the antioxidant polyphenol contained in brown bioglycerol ensures prolonged anticorrosive properties.
>
> These are properties which can have a huge impact if considered on new huge concrete structures, such as the enlarged Panama's Channel, dams, bridges, airport pavements and highways.

The discovery led to vigorous research efforts in many countries worldwide. The results were shortly confirmed and expanded in several laboratories in Panama [18], Indonesia [19], Portugal [20], and in the United States [21].

The development of greener and more sustainable cement [22], indeed, is a global issue actively researched in both academy and industry since at least two decades [23].

A few figures render the impressive growth of cement industry due to global urbanization. Production increased more than sevenfold to one billion tonnes in the 33 years between 1950 and 1983, before hitting 2 billion tonnes in 2004, and 4 billion tonnes in 2013. In 2014 around 4.2 billion tonnes of cement were produced [24].

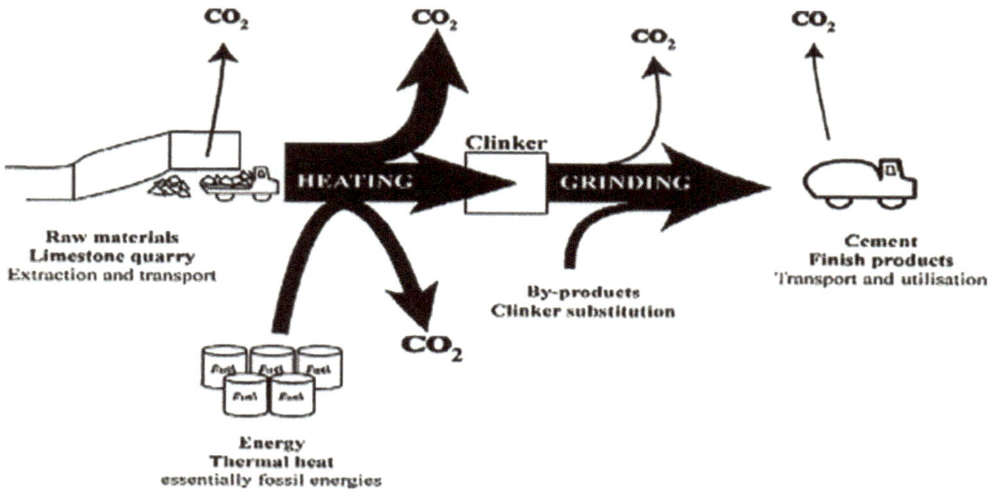

FIGURE 4.2

Following heating of the raw mix and formation of clinker, gypsum and clinker are finely grounded giving pure cement to which admixtures (secondary additives) are usually added to enhance the cement properties. *Reproduced from reference M. B. Ali, R. Saidur, M. S. Hossain, A review on emission analysis in cement industries, Renew. Sustain. Energy Rev. 2011, 15, 2252–2261, with kind permission [26].*

Cement is manufactured by heating limestone ($CaCO_3$), clay or shale (aluminosilicate), sand (silica), and miscellaneous iron oxides to about 1600°C to form the clinker mixture. The resulting cement is a mixture of various minerals whose predominant (50%–70%) mineral phase is tricalcium silicate (abbreviated as C_3S: Ca_3SiO_5) followed by dicalcium silicate (C_2S, 15%–30%), tricalcium aluminate (C_3A, 5%–10%), and ferrite (C_4AF, 10%–15%) [25].

This energy intensive process is followed by a grinding process (Fig. 4.2) in which the cement clinker is mixed with gypsum and ground in a ball mill to a finely divided state, giving a particularly high surface area to the resulting "Portland" hydraulic cement whose comprising particles are 90% (by mass) finer than 44 μm.

Typically a low-cost fuel, used to generate the large amount of thermal energy, is employed to heat the calcination furnace, whereas electrical energy is used to grind the resulting clinker in a rotating mill, and to cool the final cement product. The calcination and grinding processes account for 80% of the energy consumption, while power for the cooler accounts for the remaining 20%.

In industrially developed countries, where the cost of energy accounts for 40% of the overall cost of production, thermal energy consumption has been reduced almost to the theoretical minimum of 3500 MJ/tonne of clinker. On the other hand the consumption of power used in the grinding process comprised in the range 90–130 KWh/tonne of cement, is increasing due to growing demand for higher grade cements (increased fineness and reduced packset time).

4.4 CEMENT GRINDING AND ANTICORROSION AID

Cement grinding aids (CGAs) are organic compounds added to the clinker in the cement mill whose main purpose is to reduce the energy required to grind the clinker into a given fineness (increasing the efficiency of the cement mill). After grinding, CGAs also improve flowability and ease storage by reducing packset time and preventing hang-up to silos inner surface.

CGAs considerably decrease electrical energy consumption due to increased grindability to the required particle size (typically, 32 μm) relative to reference cement [27]. Accordingly, since more than 50 years the cement industry adds small amounts of grinding additives traditionally comprised of triethanolamine (TEA) or diethylene glycol (DEG) during the milling process [28].

The optimum dosage is achieved when a continuous monolayer is formed onto the particle surfaces affording monomolecular coverage [29], which typically requires a 0.01%–0.2% by weight of organic compound, usually added as aqueous solution. Higher amounts lead to lubrication effects with consequent decrease in the rupture of the particles and reduction in efficiency of the mill.

Crude glycerol from biodiesel streams employed as CGA is not only superior to traditionally employed grinding aids such as TEA or DEG, but it also outperforms pure glycerol as CGA, affording cement with improved mechanical and textural properties, with lower energy consumption during grinding of different clinker samples originating from widely different cement plants (two in Italy and another in Greece, Table 4.1).

To gain insight into the excellent CGA properties of biodiesel-derived, we and Michele Rossi's team studied the interaction of glycerol with cement clinkers [30].

Mixing the clinker with crude biodiesel glycerol does not induce any important chemical transformations in the structure of the cement precursor (Fig. 4.3). The main characteristic bands of the clinker material in the region 400–1000/cm are present in the mixed material. The new bands that appear in the IR spectrum are due to bioglycerol.

No significant change in the composition of the clinker in the presence of the additive was observed, even if high loadings of additive were used. The formation of polyglycerols or dehydrated/oligomeric species, which in

Table 4.1 Properties of Clinkers With Added Crude Glycerol (GLY), Triethanolamine (TEA), or Diethylene glycol (DEG)

Cement and Additive	Particle Size (μm)[a]				Blaine (cm^2/g)[b]	NMR (s)[c]	Pack Set (s)[d]
	R32	R45	R63	R90			
Italy	71.2	82.4	91.7	97.8	3580	2200	41
Italy + TEA	73.8	85.3	93.8	98.6	3510	2250	29
Italy + DEG	74.1	84.8	93.3	98.5	3590	2200	28
Italy + GLY	79.9	89.8	96.3	99.5	3590	2050	32
Greece	80.0	89.7	96.3	99.5	3610	2350	68
Greece + TEA	77.5	88.1	95.4	99.3	3510	2400	21
Greece + DEG	78.5	89.0	96.0	99.4	3620	2200	38
Greece + GLY	77.3	88.1	95.5	99.3	3530	2400	32

[a]Blaine: Test for determining the fineness of cement on the basis of its permeability to air. The value increases as the fineness increases.
[b]Laser postsource decay: Granulometry distribution (%) of particles with diameters greater than 32, 45, 63, and 90 μm.
[c]Total number of mill revolutions.
[d]Time required to break cement agglomeration under a standard test.
Reproduced from reference M. Rossi, C. Della Pina, M. Pagliaro, R. Ciriminna, P. Forni, Greening the construction industry: enhancing the performance of cements by adding bioglycerol, ChemSusChem, 1 (2008) 809–812, with kind permission.

principle can take place due to local overheating in ball mills as calcium hydroxide is a known excellent catalyst for linear polyglycerol formation, was not significant and therefore it is not responsible for the significant enhancement of the milling process due to bioglycerol.

The results of these investigations, in other words, point to surface tension modification of the clinker particles as the main effect of crude bioglycerol during the grinding process. Most of the beneficial properties of crude glycerol as CGA can be ascribed to the reduction of the agglomeration energy of cement particles as cleavage and agglomeration of cement particles play a key role in cement grinding, hardening, and ultimately, in the stability of concrete building structures.

Crude glycerol molecules behave as surfactants, reducing the energy needed to break down the particles. The two main effects significantly impacting the performance of the grinding aid molecules, indeed, are the reduction of surface polarity; and the reduction of surface energy that goes along with the reduction of the surface polarity. The high content of residual soap explains the enhanced action of crude glycerol as CGA. We remind that the crude glycerol stream from a biodiesel plant normally contains soap (free fatty acid salts) formed via undesired saponification reaction during the biodiesel production process [31].

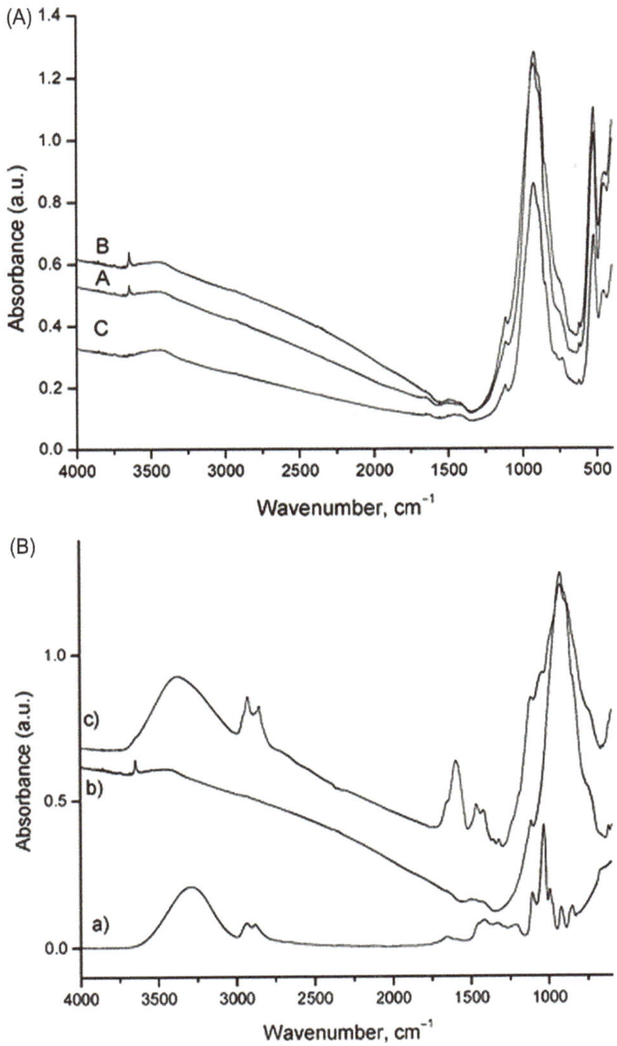

FIGURE 4.3

(A) FT-IR spectra of the three clinker samples: A: is clinker A, B: is clinker B, and C: is clinker C; (B) FT-IR spectra of a: bioglycerol, b: clinker B, and c: clinker B + 10 wt% bioglycerol. *Reproduced from reference A. Parvulescu, M. Rossi, C. Della Pina, R. Ciriminna, M. Pagliaro, Investigation of glycerol polymerization in the clinker grinding process, Green Chem. 13 (2011) 143–148, with kind permission.*

Admixtures interact with the cement constituents and affect the properties of fresh and hardened concrete. Recently Heinz and Mishra published a theoretical study [32] suggesting that the main mode of action of grinding aid polar molecules, including glycerol, lies in the reduction of agglomeration energy. The study, furthermore, was able to disprove the role of electrostatic repulsion as no evidence of significant quantities of charges on clinker surfaces was found (ionic grinding aids, e.g., do not have advantages over neutral ones).

Results suggest that the grinding aid molecules turn their polar functional groups toward the clinker surface. In detail the polar hydroxyl groups (-OH) of the grinding aid reduce the surface polarity already partially offset by the soap molecules at the surface of the clinker. The surface-bound glycerol molecules act as a spacer to reduce agglomeration of the highly polar surfaces by creating an organic interlayer that effectively mitigates the attraction between the charged surfaces. As a result clinker surfaces coated with a glycerol monolayer attract each other moderately, enabling formation of smaller agglomerates and enhancing the grinding efficiency; whereas, in comparison to tertiary amino alcohols, the distance of glycerol to superficial ions is smaller, resulting in stronger binding.

The anticorrosion properties of crude glycerol explain why the addition of crude glycerol enhances the resistance to corrosion of finished concrete structures, despite the presence of chloride.

Crude glycerol with its substantial color (yellow to dark brown) has a remarkably high α-tocopherol content (0.04–6 mmol/L), which leads to the excellent anticorrosive and antioxidant properties discovered by Jerzykiewicz and coworkers in Poland [33]. In detail the antioxidants present in the glycerol fractions from biodiesel production have an α-tocopherol-like structure, and their antioxidant activity is due not only to the phenolic groups, but to the tocopherol-like structure itself as the adducts formed from carbon-centered radicals of α-tocopherol (following radical attack at several α-tocopherol positions, Fig. 4.4) are energetically favored in relation to the phenolic ones.

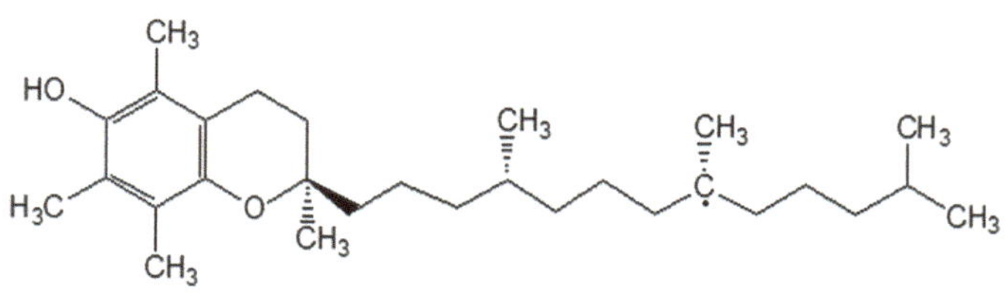

FIGURE 4.4

Carbon-centered radical formed upon radical attack of a radical on a nonphenolic position. *Reproduced from reference M. Jerzykiewicz, I. Ćwieląg-Piasecka, M. Witwicki, A. Jezierski, EPR spin trapping and DFT studies on structure of active antioxidants in biogycerol. Chem. Phys. Lett. 497 (2010) 135–141, with kind permission.*

4.5 CEMENT STRENGTH ENHANCER AND WATERPROOFING AGENT

Usually added at the clinker milling step, a variety of additives are used in industry as curing agents to improve the resistance to fracture of the finished cement. High compression strength of cement structures is obviously important. Curing agents prevent fast hardening of wet cement paste, which decreases its resistance to compression. They work by preventing fast evaporation of water, and are usually comprised of polyols and other hydrophilic molecules. The high cost of refined glycerol, a well-known effective agent capable to improve compression strength [34], has traditionally limited its use in the cement industry for decades. Low-cost crude glycerol is a better curing agent, when compared to pure glycerol, with optimal addition of 400 ppm of crude glycerol affording between 5% and 7% in the resistance to compression of the concrete object.

Table 4.2 lists the results of compressive tests conforming to the European standard EN 196 ("Methods of testing cement") and related parts ("Part 1: Determination of strength"; "Part 6: Determination of fineness"), carried out on two types of commercial clinker. The performance of crude glycerol produced by a biodiesel plant is compared with that of high-purity-grade glycerol. The cement precursor was blended with gypsum and milled for a given time in order to produce cements with a similar fineness (Belgium, Greece, and Italy are arbitrary names of different types of cement)

Table 4.2 Compression Strength of Three Cement Clinkers With Addition of Pure or Crude Glycerol[a]

		Particle Size (μm)[c]				Compressive Strength (MPa)			
Cement and Additive	Blaine (cm^2/g)[b]	R32	R45	R63	R90	1 day	2 days	7 days	28 days
Belgium	3230	21.9	10.5	3.2	0.1	n.d.	25.9	45.0	58.2
Belgium + pure glycerol	3290	24.7	12.5	4.1	0.4	n.d.	26.7	45.0	58.8
Belgium + crude glycerol	3160	26.5	14.2	5.2	0.7	n.d.	28.3	46.9	60.8
Greece	3570	18.8	8.7	2.6	0.2	16.1	n.d.	41.6	53.0
Greece + pure glycerol	3550	22.1	10.8	3.4	0.3	20.1	n.d.	41.6	53.8
Greece + crude glycerol	3590	21.5	10.7	3.6	0.5	18.1	n.d.	43.8	56.6

[a]*Standard procedures such as those described in the European standard EN 196 were used to evaluate the properties of the clinkers herein. Analytical data for crude glycerol (density = 1.29 kg/L) are as follows (wt %): glycerol 92 %; water 1.5 %; NaCl 5 %; nonglycerol organic matter 0.8 %.*

[b]*Blaine: Test for determining the fineness of cement on the basis of its permeability to air. The value increases as the fineness increases.*

[c]*Laser postsource decay: Granulometry distribution (%) of particles with diameters greater than 32, 45, 63, and 90 μm.*

Reproduced from reference M. Rossi, C. Della Pina, M. Pagliaro, R. Ciriminna, P. Forni, Greening the construction industry: enhancing the performance of cements by adding bioglycerol, ChemSusChem 1 (2008) 809–812, with kind permission.

in the presence of crude or pure glycerol, which was added as 50 % solutions in water in amounts comparable to those used with current commercial additives such that the final concentration was 400 ppm.

In all cases crude glycerol (containing 5 % NaCl and with an amber-brown color) was found to give improved mechanical properties with respect to those produced by pure glycerol.

In 2013 Sebastián Vásquez Bonilla and coworkers in Panama confirmed and extended the results above [18].

Concrete cylinders and joists were prepared by filling metal cylinders or joists with wet concrete. After 45 min from pouring the cement paste, a thin layer of crude glycerol from a Panama's biodiesel plant was spread out on the surface of the set concrete mixtures by repeating at regular intervals three times during the solar day (in the morning, midday, and in the afternoon) the addition of crude glycerol (Fig. 4.5).

Resistance to compression of the hardened concrete structures for both joist and cylinder samples obtained after 7 and 14 days of curing was measured according to the ASTM 39/C 39M05 (for the compression of cylinders) and ASTM C78-02 (for the flexion of small joists) standards.

Results in Table 4.3 shows that the resistance to compression was between 10% and 15% higher for the joist structures made with bioglycerol added cement, and 2.7%–10% higher for the corresponding cylinders pointing, as expected, to insufficient penetration of crude glycerol in the relatively high cylinder cement body.

FIGURE 4.5

Cylinders of concrete midday (Left) and at the end of the same day (Right). The slightly colored cylinders are the samples added with crude glycerol. *Photograph of Prof. S. Vásquez Bonilla, reproduced with kind permission.*

Table 4.3 Resistance of Concrete Objects With and Without Crude Bioglycerol. For Cylinders the Resistance Toward Compression and for Small Joists the Resistance to Flexion Was Measured

Cement Object Shape	Sample	(kN)	(N/mm^2)
	7 days		
Small joist	Reference	13.89	1.79
Small joist	Reference + crude glycerol	16.11	2.08
Cylinder	Reference	85.30	10.36
Cylinder	Reference + crude glycerol	89.90	10.92
	14 days		
Small joist	Reference	17.89	2.31
Small joist	Reference + crude glycerol	19.83	2.56
Cylinder	Reference	92.90	11.46
Cylinder	Reference + crude glycerol	95.50	11.77

Reproduced from reference S. Johnson, S. Vasquez Bonilla, N. Yao, Valorización de la glicerina cruda obtenida como producto secundario en la producción de biodiesel. Tecnociencias 15 (2013) 41–55, with kind permission.

Recently researchers in Portugal confirmed and extended results found in Italy and in Panama showing further evidence that crude glycerol behaves even better than commercial accelerating admixtures and antifreezing admixtures, at a fraction of the cost [35].

In detail the team confirmed the role of crude glycerol at optimum 1 wt% concentration (beyond which saturation occurs) acting as an accelerator admixture, both for setting and general hydration kinetics. Incorporating biodiesel crude glycerol in cement and mortars, they found that the inclusion of glycerol caused (1) the reduction of the initial and final setting times of the cement paste samples, and (2) a significant increase in the kinetics of stiffness and strength evolution in the cement paste samples since early ages.

Crude glycerol behaves similarly to reagent grade pure glycerol (cement paste mixes *P1%g* and *P1%rg*) affording a material with higher *E*-modulus (11.8 GPa for the mix 1% glycerol compared to 11.6 GPa for the reference mix), but at much faster pace than the reference, without impacting negatively the development of this fundamentally important mechanical property after the advantageous effect of acceleration has finished.

The researchers led by Miguel Azenha discovered a dramatic reduction of permeability properties in mortars (both to water and oxygen), which indicates that this material can be considered for durability improvement; and a slight reduction of shrinkage of cement pastes containing 0.5% and 1% of crude glycerol. The oxygen permeability in mortar aged 28 days containing 0.5% crude glycerol is 2.76 lower than that of the reference mortar, whereas the cement paste added with 1% crude glycerol has 3.5 lower water permeability.

The glycerol and soap molecules present in crude biodiesel glycerol, in other words, partly fill the cement micropores, as it happens in the interstices between cell living tissues wherein glycerol is used to prevent frosting-associated cell death.

4.6 FORM RELEASE AGENT

Vásquez Bonilla and coworkers in Panama discovered that crude biodiesel glycerol added to cement is an effective form releasing agent allowing easier removal of mold structures used during construction [18]. Fig. 4.6 shows evidence that easy removal of the molds from the concrete structures made from cement modified with crude glycerol leaves the molds clean, whereas no stains are formed on the concrete structures as it happens often with mineral oils.

In detail crude bioglycerol was spread out onto the internal surfaces of several molds of cylinders and joists. A freshly prepared fluid concrete mixture was then poured in the molds. After 48 h the molds were stripped and the clean, excellent state of the inner surface was evident. The molds were easily washed with water, and left ready for reuse.

FIGURE 4.6

Molds and cylinders with cement residues following stripping of the set cement after 48 h. The mold in front, containing cement added with crude glycerol, is clean. *Photograph of Prof. S. Vásquez Bonilla, reproduced with kind permission.*

In the concrete construction industry, form release agents prevent the adhesion of freshly placed concrete to the forming surface usually made of plywood, steel or aluminum. A good release agent promotes the clean release of forms and helps to reduce surface imperfections in concrete, minimizing dusting of formed surfaces and reducing labor costs associated with stripping and cleaning forms and equipment. Traditionally form coatings are lubricant oils derived from petroleum, including diesel fuel and used motor oil [36]. Typically during construction with plywood panels, the panels are treated with the chemical release agent before the first use, and between each pour. This will prolong the life of the plywood panel by enhancing its release characteristics.

To explain this additional excellent property of crude glycerol of interest to the construction industry, it may be noted that in case of fresh concrete, hydration reaction will dominate over partial hydroxylation reaction due to presence of enough amount of water. Hence, the glycerol molecules in crude bioglycerol react with hydrated tricalcium silicate C_3S particles in hydrated concrete to form a soapy film which prevents adhesion, leaving no residue on the forming concrete surface. The presence of soap molecules in crude glycerol further enhances such anti-adhesion action, leading to better form release agent. Furthermore glycerol in crude bioglycerol is an effective antifreeze agent, which favors its preferential use in cold weather conditions.

4.7 THE BUSINESS CASE FOR BIOGLYCEROL AS CEMENT AID

Today raw glycerol by-product of biodiesel manufacturing is a performance enhancer of cement, increasingly employed by the concrete construction industry. Already in 2010 when the annual market for grinding aids in the EU exceeded 50,000 tonnes, market consumption of crude glycerol as CGA was estimated to be at least 10,000 tonnes.

A decade ago a large construction products company started to commercialize crude glycerol as cement quality improver, first in Europe and then worldwide [37]. The new additive, and its reliable sourcing from one or more selected biodiesel plants, proved to be capable to effectively replace oil-derived additives purchased by a few large oil refineries. For example, during the devastation brought about by hurricane Katrina in 2005 in the United States, the New Orleans petrochemical refineries were shut down, interrupting supply of numerous chemicals, including ethylene and propylene glycols. This led the above construction products company to replace oil-derived glycols with crude glycerol from biodiesel refineries.

Since then, the commercial utilization of crude glycerol by the construction industry has steadily increased, even though figures are generally not published. What is remarkable from a scientific and environmental viewpoint, is that the use of crude glycerol in construction provides additional environmental benefits that should not be underestimated. In the United States, e.g., the use of diesel fuel as form releasing agent has recently been banned. Diesel oil is combustible, contributes to smog and presents health and safety risks to workers. Both diesel and used motor oils also contaminate the soil and groundwater where the cement was unloaded and applied. New bioglycerol-based cement release products,[a] on the other hand, are comprised of biodegradable crude glycerol thereby fully meeting the new environmental regulation.

Construction companies using crude glycerol, in their turn, will benefit with better, faster construction process with reduced occupational health hazards and a far better environmental "footprint" of their products: the buildings in which we all live and work.

References

[1] L.B. Lane, Freezing points of glycerol and its aqueous solutions, Ind. Eng. Chem 17 (1925) 924.

[2] C.S. Miner, N.N. Dalton, Glycerol, Reinhold Publishing Company, New York, 1953.

[3] K.-i Murata, H. Tanaka, Liquid–liquid transition without macroscopic phase separation in a water–glycerol mixture, Nat. Mater 11 (2012) 436–443.

[4] Å. Melinder, Thermophysical Properties of Aqueous Solutions Used as Secondary Working Fluids, Royal Institute of Technology, KTH, Stockholm, 2007, <www.diva-portal.org/smash/get/diva2:12169/FULLTEXT01.pdf>.

[5] J. Greenwell, PG to Glycerin—Lessons Learned, energy.gov, May 5, 2015.

[6] M.A. Lesser, Glycerin—Man's most versatile chemical servant, J. Chem. Educ. 26 (1949) 327–328.

[7] B. Ortner, Glycerine Won't Replace MEG Say Anti-freeze Blenders, icis.com, March 2, 2006.

[8] R. Hudgens, R. Hercamp, J. Francis, D. Nyman, Y. Bartoli, An evaluation of glycerin (glycerol) as a heavy duty engine antifreeze/coolant base, SAE Technical Paper 2007-01-4000, 2007.

[9] ASTM International, Proposed ASTM Engine Coolant Standards Focus on Glycerin, Press Release #8555, April 2010.

[10] Lucintel, Global Automotive Coolant Market 2014-2019: Trends, Forecast, and Opportunity Analysis, New York, 2014.

[11] S.M. Roberts, R.C. James, P.L. Williams, Principles of Toxicology: Environmental and Industrial Applications, Wiley, New York, 2015.

[12] Grand View Research, U.S. Automotive Coolant Market Analysis by Product (Ethylene Glycol, Propylene Glycol, Glycerin), by Technology (Inorganic Acid, Organic Acid, Hybrid Organic Acid), by Distribution (OEM, Automotive Aftermarket), by End-Use (Passenger Car, Commercial Vehicles, Two Wheelers) and Segment Forecasts to 2024, San Francisco, August 2016.

[a]See, e.g., the *Eco Slide* product developed in 2013 by Extreme Biodiesel in the United States.

[13] M. Rossi, M. Pagliaro, R. Ciriminna, C. Della Pina, W. Kesber, P. Forni, Improved Compression Strength Cement, WO2006051574, (2006).

[14] M. Rossi, C. Della Pina, M. Pagliaro, R. Ciriminna, P. Forni, Greening the construction industry: enhancing the performance of cements by adding bioglycerol, ChemSusChem 1 (2008) 809–812.

[15] S.U. Khan, M.F. Nuruddin, T. Ayub, N. Shafiq, Effects of different mineral admixtures on the properties of fresh concrete, Sci. World J. 2014 (2014) 1–11.

[16] Jayant D. Bapat, Mineral Admixtures in Cement and Concrete, CRC Press, Boca Raton, 2012.

[17] C. Piccirillo, Bioglycerol and Its Use in the Construction Industry, decodedscience.org, April 17, 2015.

[18] S. Johnson, S. Vasquez Bonilla, N. Yao, Valorización de la glicerina cruda obtenida como producto secundario en la producción de biodiesel, Tecnociencias 15 (2013) 41–55.

[19] O. Farobie, S.S. Achmadi, L.K. Darusman, World academy of science, Eng. Technol. 6 (2012) 793–798.

[20] I. Carlos-Alves, M. Azenha, C. Lucas, J. Granja, Estudo experimental da viabilidade de utilização de sub-produtos do biodiesel como aditivos para materiais cimentícios, *JPEE 2014*, Lisboa, November 26–28, 2014.

[21] B. Tran, S. Bhattacharja, Bioglycerol By-Products and Methods of Using Same, US 7892353 B2.

[22] I. Amato, Green cement: concrete solutions, Nature 494 (2013) 300–301.

[23] M. Schneider, M. Romer, M. Tschudin, H. Bolio, Sustainable cement production—present and future, Cem. Concr. Res. 41 (2011) 642–650.

[24] H.G. Van Oss, USGS, US Department of the Interior, Mineral Commodity Summaries 2015, <minerals.usgs.gov/minerals/pubs/mcs/2015/mcs2015.pdf>.

[25] H.F.W. Taylor, Cement Chemistry, second ed., Academic Press, London, 1997.

[26] M.B. Ali, R. Saidur, M.S. Hossain, A review on emission analysis in cement industries, Renew. Sustain. Energy Rev. 15 (2011) 2252–2261.

[27] I. Teoreanu, G. Guslicov, Mechanisms and effects of additives from the dihydroxy-compound class on Portland cement grinding, Cem. Concr. Res. 29 (1999) 9–15.

[28] C.J. Engelsen, Quality improvers in cement making—State of the art, COIN Project P1 Advanced Cementing Materials, SINTEF Building and Infrastructure, Oslo, 2008.

[29] M. Weibel, R.K. Mishra, Comprehensive understanding of grinding aids, ZKG 6 (2014) 28–39.

[30] A. Parvulescu, M. Rossi, C. Della Pina, R. Ciriminna, M. Pagliaro, Investigation of glycerol polymerization in the clinker grinding process, Green Chem. 13 (2011) 143–148.

[31] S. Hu, X. Luo, C. Wan, Y. Li, Characterization of crude glycerol from biodiesel plants, J. Agric. Food Chem. 60 (2012) 5915–5921.

[32] R.K. Mishra, R.J. Flatt, H. Heinz, Force field for tricalcium silicate and insight into nanoscale properties: cleavage, initial hydration, and adsorption of organic molecules, J. Phys. Chem. C 117 (2013) 10417–10432.

[33] M. Jerzykiewicz, I. Ćwieląg-Piasecka, M. Witwicki, A. Jezierski, EPR spin trapping and DFT studies on structure of active antioxidants in biogycerol, Chem. Phys. Lett. 497 (2010) 135–141.

[34] H. H. Moorer, C. M. Anderegg, Cement Grinding Aid and Set Retarder. US 4204877 A (1979).

[35] M. Azenha, C. Lucas, J. Luís Granja, I. Carlos-Alves, E. Guimarães, Glycerol resulting from biodiesel production as an admixture for cement-based materials: an experimental study, Eur. J. Environ. Civil Eng. (2016), 1–17. <http://dx.doi.org/10.1080/19648189.2016.1177603>.

[36] M.K. Hurd, Contractor's guide to form release agents, Concrete Construction, September 1999.

[37] M. Pagliaro, M. Rossi, The Future of Glycerol, RSC Publishing, Cambridge, 2010, Chapter 11.

CHAPTER 5

Glycerol: A Key Platform Chemical of the Forthcoming Bioeconomy

5.1 BIO-BASED PRODUCTS AND LOW OIL PRICES

All bioplastics and all bioderived chemicals are held back by low oil price and fierce competition from nonrenewable and nonrecyclable plastics made from low cost oil. The same holds true for any product obtainable from bioglycerol, with the exception of highly valued derivatives such as polyglycerols, advanced functional polymers, and other specialty chemicals.

As put it by Cascone, principal at a leading chemical market consultancy (Nexant), history has shown that the development of bioplastics such as polylactides (PLAs) and polyhydroxyalkanoates (PHAs) has been slow and problematic [1].

Hence the answer whether or not chemicals and polymers can be conveniently derived from bioglycerol is part of the answer to the question whether a bioeconomy, namely the switch in the production of goods and energy from fossil to renewable energy and material resources (biomass) [2], is truly feasible within a realistically short time frame.

This inevitably leads to the issue of oil price and availability.

This chapter investigates the issue, following a recent study [3] conducted by us along with Francesco Meneguzzo's team at Italy's Research Council in which the competing dynamics of oil availability, demography, and economic growth were combined for the first time in the frame of modern nonlinear scaling theories.

In the course of the last 15 years (2000–15), one of the most impressive changes in the energy scenario has been the accelerated deployment of renewable energy sources (RES) at the global scale, so much that the perspective of worldwide wind and solar electricity supply is no longer a futile exercise [4]. Along with wind energy, easily scalable and versatile solar photovoltaics (PV) has been massively adopted, first in Europe and now

Glycerol. DOI: http://dx.doi.org/10.1016/B978-0-12-812205-1.00005-9

across the world, to become the energy technology with the fastest rate of adoption, and chances for further improvement [5].

Nevertheless, since 2009 another large change occurred in the exploitation of new energy sources, namely, the successful extraction of unconventional hydrocarbons, mainly shale gas and shale oil (the latter hereinafter referred to as "tight oil"), relying on a combination of new and massively deployed extraction techniques, such as horizontal drilling and fracking.

Started to be seriously explored in the United States in 2007–08, such resources passed into massive production in mid-2011, allowing that country to attain self-sufficiency in gas production in a dramatically short lapse of time [6].

In detail more than 4 million barrels per day of tight oil added to the global production of petroleum (roughly 5% of the world's production), with the most noticeable increase occurring after 2010, driven by sustained high oil prices [7]. Eventually in 2015 the global oil supply had increased by more than additional 9 million barrels per day, challenging the very concept of "peak oil," namely the theory predicting that the world's production of crude oil should have already reached a maximum (peak), to gradually decline until vanishingly levels [8].

In the spring of 2015, however, a possible imminent decline of tight oil successfully extracted during the previous 4 years was announced [9]. Early signs of production decline arose in March 2015 weekly data [10], following the steep fall of the oil rigs count since November 2014 [11], while the oil price per barrel was falling from about $110/barrel on June 2014, to about $48/barrel in January 2015.

Certain energy analysts have argued that, once the oil price will rebound toward $100/barrel, the uncompleted wells will be resumed, and oil rigs count shall increase again. Others, conversely, rule out any chance for production recovery in light of the demonstrated sensitivity of the tight-oil industry to the oil price [12].

Due to its intrinsically high extraction costs, the "well-head" energy return on energy invested (EROI) for tight oil is in 1.5–2 range [13], namely about one-tenth of the 11–20 set of values for conventional crude oil [14].

The technical reasons for such a low EROI are the use of large quantities of energy from the same shale formation during the process (including liquefaction of the originally solid kerogen), as well as the need for a huge number of drills, thus of oil rigs, in order to keep pace with the steep decline rates of single wells [15].

5.2 POPULATION, WEALTH, AND ENERGY

Following nonlinear scaling theories recently introduced in socioeconomic studies, [16] the world gross domestic product (GDP) was modeled as a power function of the global population. Data including demographic, energetic, and financial information at the annual and monthly frequency, at the geographical scale of the whole world as well as of single countries, were sourced from reputable sources (Table 5.1).

Fig. 5.1 shows the power law fitting the world GDP data series, plotted against the population, formally represented by Eq. 5.1, which explains more than 99% of the sample variance:

$$\mathrm{W\_GDP} = (0.0077 \pm 0.0008)\cdot \mathrm{W\_Pop}^{(4.69 \pm 0.06)} \quad (5.1)$$

where W_GDP is the world GDP, in units of trillion current US dollars, and W_Pop is the global population, in units of billions.

Simple differentiation of Eq. 5.1 provides the expected year-on-year GDP growth rate d(W_GDP) as described by Eq. 5.2, where d(P) is the population growth rate, in the same units as Eq. 5.1:

$$d(W\_GDP) = 0.0361\cdot \mathrm{W\_Pop}^{3.69}\cdot d(P) \quad (5.2)$$

Similar relationships (not shown) such as the one described by Eq. 5.1 can be easily proved to hold for specific economies and single countries, such as European Union, Germany, France, Italy, or the United States, showing that the power law fitting of the GDP as a function of population has been an ubiquitous feature of the world's economic development at

Table 5.1 Datasets and Sources

Dataset	Period and Frequency	Unit	Source
World population	1950–2015 Annual	Billions	UN [17]
World gross domestic product (current US$)	1960–2015 Annual	Trillion US$	World Bank [18]
Broad money Industry value added	1960–2015 Annual	% of GDP	World Bank [18]
World energy consumption by source	1965–2015 Annual	MTOE	BP [19]
Brent oil price	1965–2015 Annual	$ per barrel	BP [19]
	1987–2016 (May) Monthly		EIA [10]
World and regional crude oil production	1987–2016 (May) Monthly	Million barrels per day	JODI [20]

Reproduced from reference F. Meneguzzo, R. Ciriminna, L. Albanese, M. Pagliaro, The energy-population conundrum and its possible solution, arXiv:1610.07298 [physics.soc-ph], with kind permission.

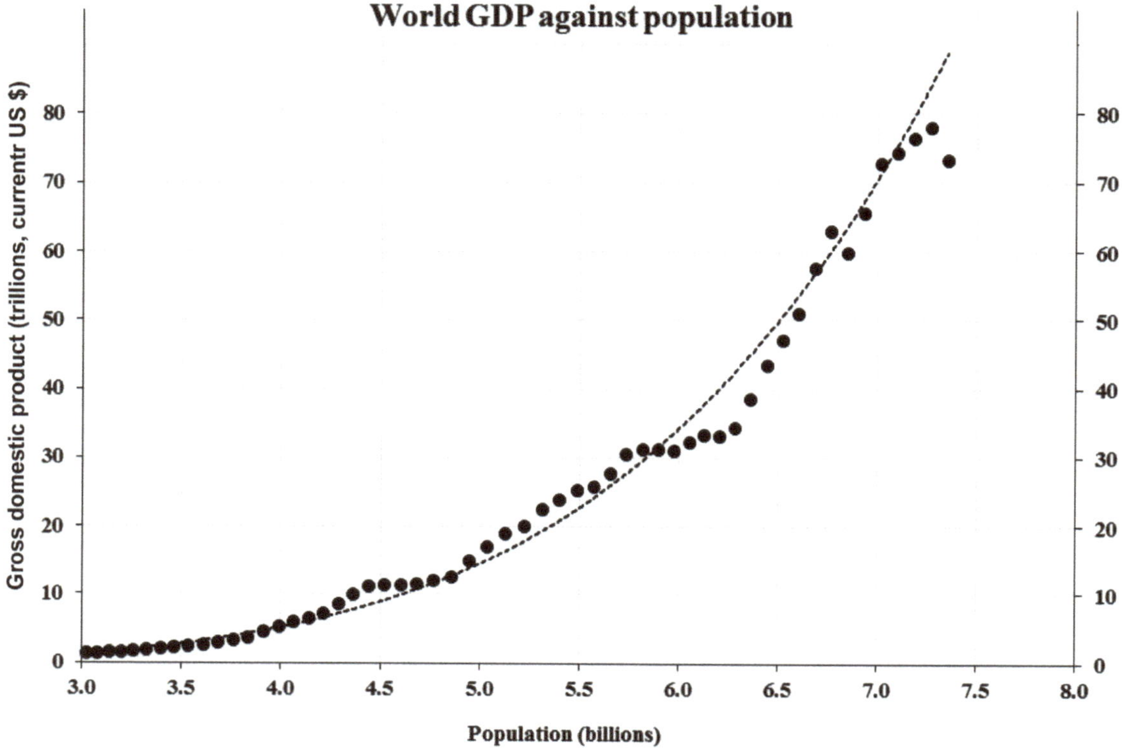

FIGURE 5.1
Power law fitting the gross domestic product (GDP) as a function of the global population, during 1960–2015. *Reproduced from reference F. Meneguzzo, R. Ciriminna, L. Albanese, M. Pagliaro, The energy-population conundrum and its possible solution, arXiv:1610.07298 [physics.soc-ph], with kind permission.*

least since 1960, with a specific exponent of the power law assigned to each economic area.

Nevertheless a deviation from the exponential relationship looks like to have arisen during the last 2 years of the series, i.e., 2014 and 2015, with a sudden downturn of GDP in 2015.

As shown in Fig. 5.2A, the total energy consumption (TEC) has steadily increased, literally fueling the growth of global wealth, although at a pace only proportional to the overall population (98% of variance explained by a linear fit).

In detail Fig. 5.2B shows that the specific (per capita) TEC has enjoyed two distinct growing periods since 1965, the first one in 1965–73, and the second during 2001–08, with GDP growth rate accelerating in the latter period and pointing to an intrinsic crucial role of the population size.

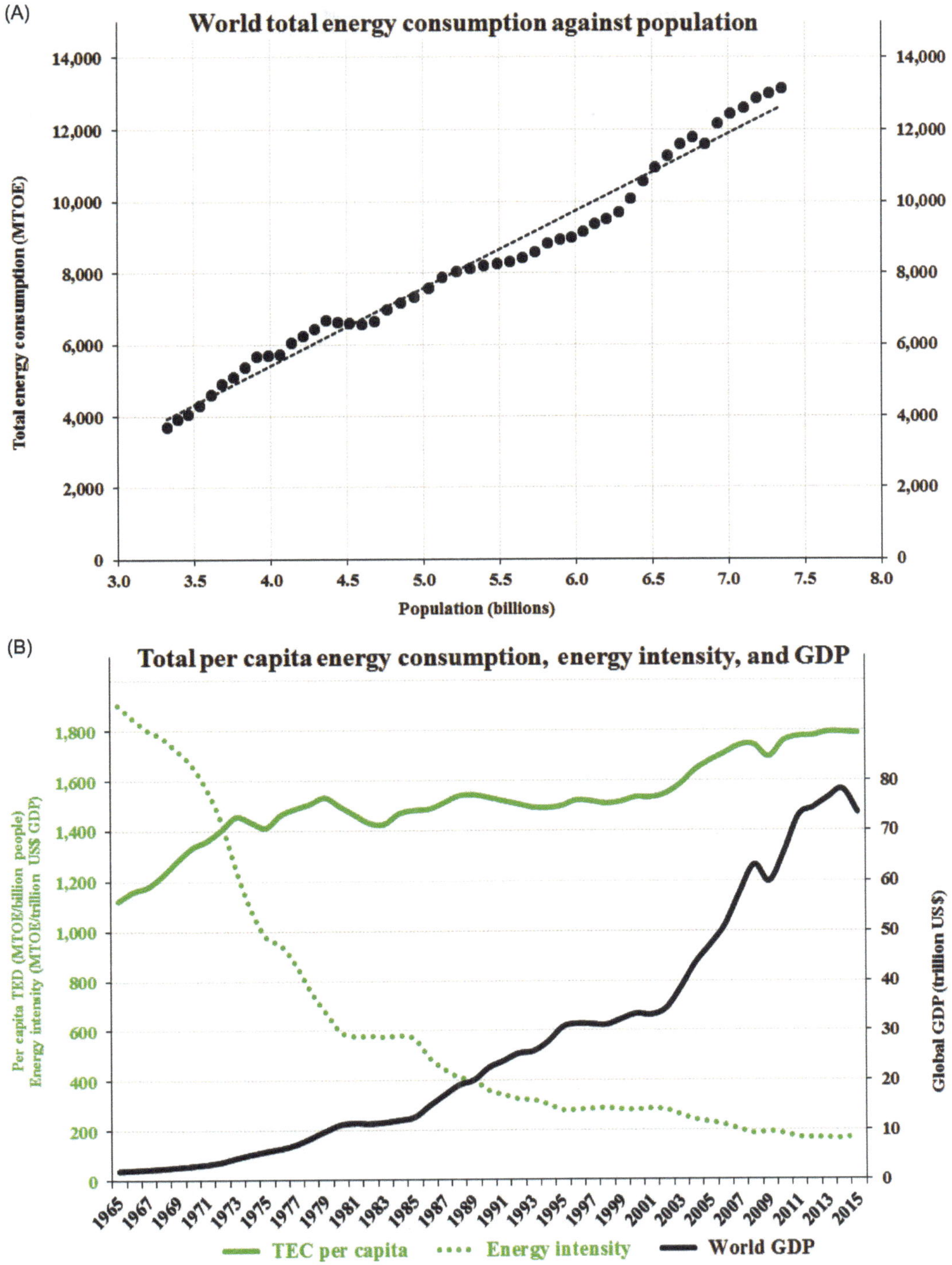

FIGURE 5.2

World total energy consumption (TEC) as a function of population (A), and time series of per capita TEC, energy intensity, and GDP (B), during the 1965–2015 period. *Reproduced from reference F. Meneguzzo, R. Ciriminna, L. Albanese, M. Pagliaro, The energy-population conundrum and its possible solution, arXiv:1610.07298 [physics.soc-ph], with kind permission.*

However, in the last 3 years of the series (2013–15), the per capita TEC was leveling off, and the GDP growth first slowed down and eventually reversed with the worrisome steep drop in year 2015.

Fig. 5.2B also shows evidence that the energy intensity (energy consumed per unit GDP) followed an impressive decline that closely resembles an exponential decay trend, although an apparent leveling is visible in the latter years of the series.

As shown in Fig. 5.3A, the energy mix has profoundly changed during the 50-years period under consideration, most notably with the gradual decline of the oil contribution, particularly since the late 1970s, partly replaced by natural gas and coal. Nuclear energy, remarkably, has covered a tiny fraction of the global consumption and, even more important, it peaked in 2006 at 635 MTOE (5.6% of the total). In 2015 the figures for the nuclear source reduced to 583.1 MTOE and 4.4%, respectively.

At the same time, the global amount of money, faithfully representing the amount of debt [21], started a relentless growth (Fig. 5.3B), accelerating just during the steeper phases of oil decline in the energy mix, i.e., during most of the 1980s and in the early 2000s, hinting to the attempt by central governments and financial institutions to buy the current growth from the future amid a tightening resource basis useful for growth [22–24].

To further support this point, recently Tverberg drew attention to the role of the energy concentration as key parameter in the hierarchy of energy sources, as well as a major driver of economic growth, arising mainly from the connection between concentration and transportability [25].

Oil is by far the most concentrated energy source applicable to virtually all end uses. Hence it may not surprise that its downward trend in the energy mix has led to substantial economic difficulties. As a clue to such difficulties, Fig. 5.3C shows that industry value added as a percentage of GDP (global data available only during 1995–2014) has indeed been falling quite rapidly.

This evidence suggests that (1) most of GDP growth, at least in the last two decades, can be traced back more to the increase in debt than to real wealth generation, and (2) the concurrent and thoroughly established effect of the declining EROI of traditional fuels has played a substantial role [26].

It may be concluded that the chance of future economic growth matching the current trajectory of the human population is inextricably bound to the wide and growing availability of highly concentrated energy sources provided with broad applicability to energy end uses.

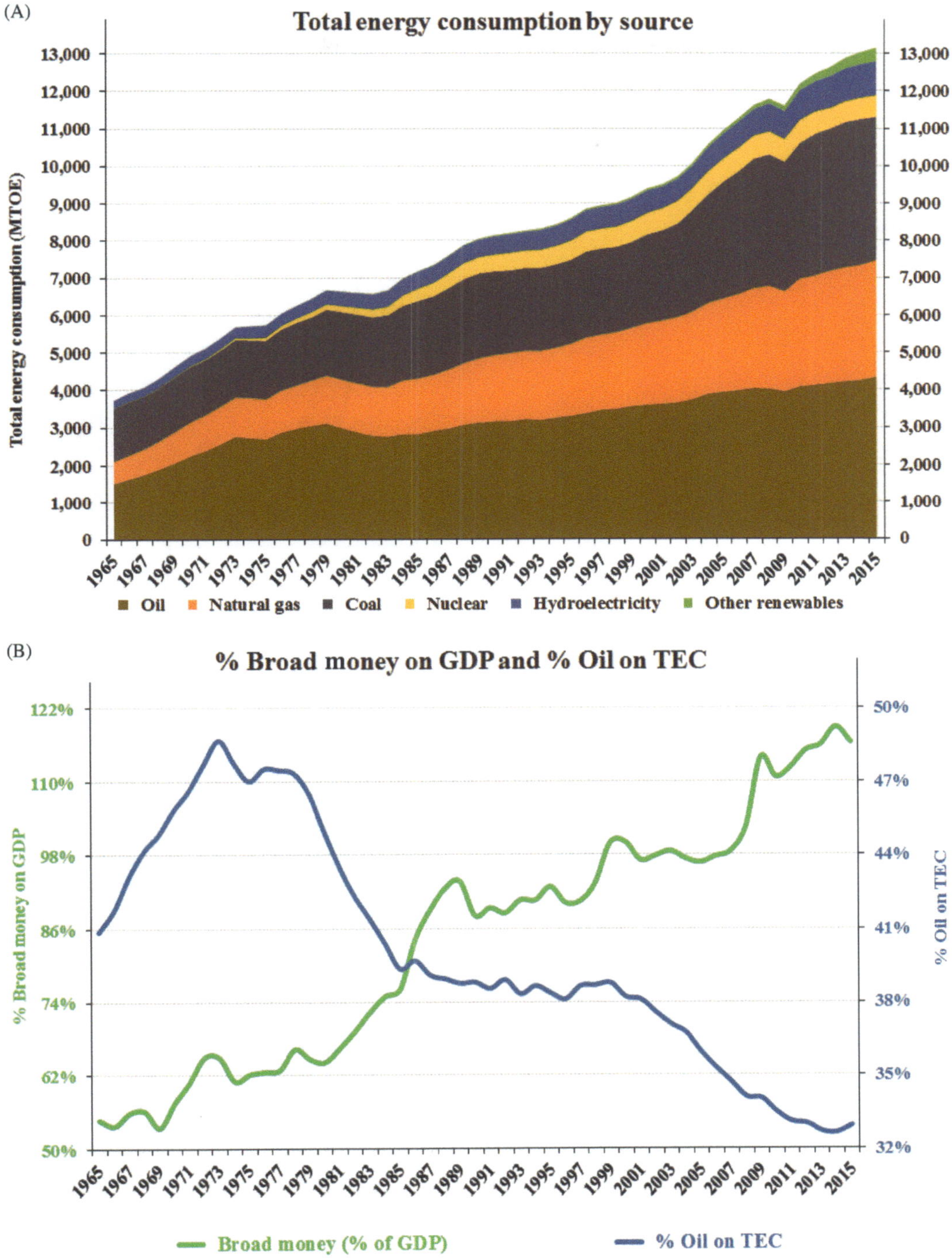

FIGURE 5.3

Energy consumption by source (A), time series of broad money and percentage oil on TEC (B), industry value added as percentage of GDP (C), during 1965–2015. *Reproduced from reference F. Meneguzzo, R. Ciriminna, L. Albanese, M. Pagliaro, The energy-population conundrum and its possible solution, arXiv:1610.07298 [physics.soc-ph], with kind permission.*

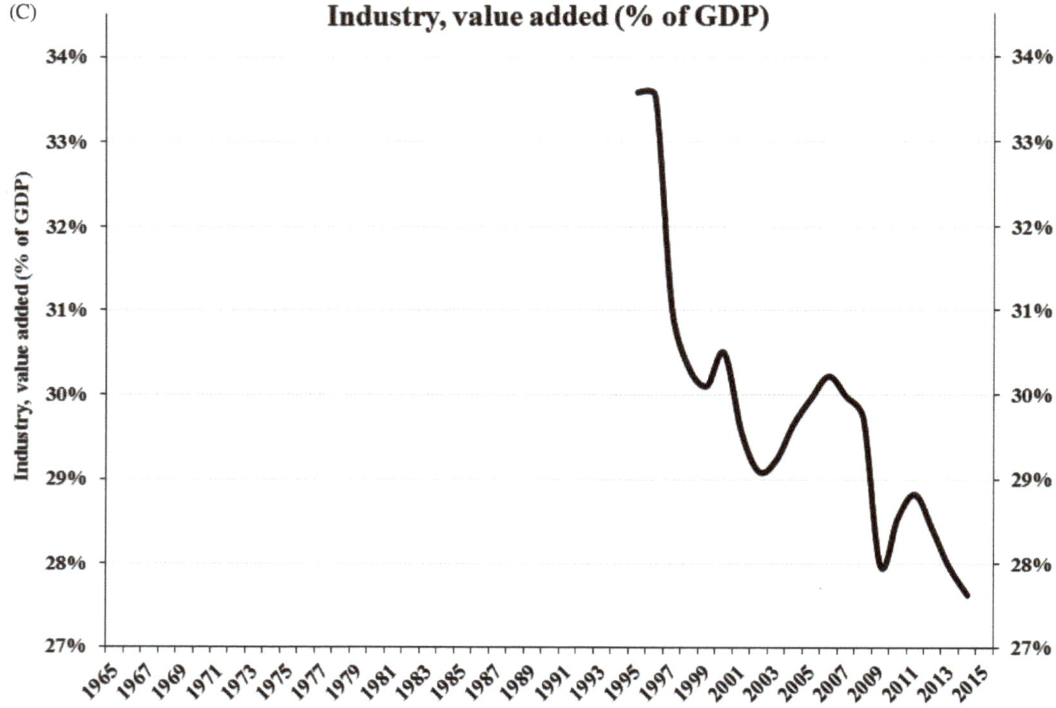

FIGURE 5.3
(Continued).

5.3 OIL VALUE AND THE GLOBAL ECONOMY

Sufficient clues exist that the oil price rising above some threshold has been a concurrent cause of recurrent slowdowns and recessions of the global economy [26,27]. In particular, spikes in oil price were deemed responsible for 10 out of 11 recessions in the United States since World War II [28]. More in general, higher cost of oil has been shown to effectively act as a "tax" inhibiting economic growth [29].

In the frame of the population-driven, energy-assisted, simple model of wealth growth shown by Eq. 5.1, Fig. 5.1, and Fig. 5.2A, an unbiased index of anomalies affecting the global economy is the difference between year-on-year growth rates of the GDP computed from real data and from the simple model described by Eq. 5.2.

The plot of such GDP growth anomalies in the period 1961–2015 along with inflation-adjusted oil price since 1955 (Fig. 5.4) shows four main events (A, B, C, and D) allegedly connecting in a causative manner the spikes in oil price with subsequent global economic slowdown.

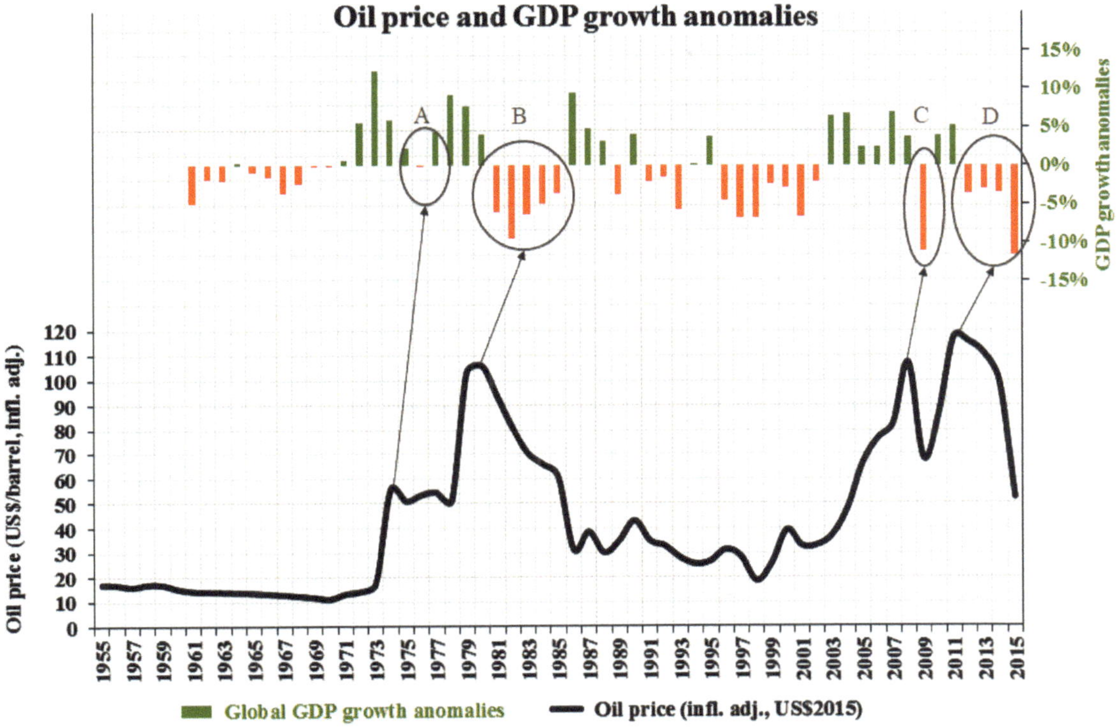

FIGURE 5.4

Time series of inflation-adjusted oil price and global GDP growth anomalies. Event labeled A–D mark economic slowdown or recession linked to high oil prices. *Reproduced from reference F. Meneguzzo, R. Ciriminna, L. Albanese, M. Pagliaro, The energy-population conundrum and its possible solution, arXiv:1610.07298 [physics.soc-ph], with kind permission.*

The "A" event occurred after the first "oil shock" in 1973–74 (maximum yearly average price about $56/barrel), whose effect was short-lived, limiting to 1976 (but occurring during a relatively strong positive deviation). The "B" event occurred after the second oil shock in 1979–80 (maximum yearly average price about $106/barrel), with a significant and sustained slowdown during the years 1981–85. The "C" event occurred immediately after the strong price spike of 2008 (average price $107/barrel), following a period of sustained growth taking place since 2003, resulting in the earnest recession of 2009.

Finally following high prices culminated in 2011 with the $117/barrel value, the "D" event featuring a series of negative anomalies of GDP growth in 2012–15, with the latter year marked by the strongest negative anomaly during the whole 1961–2015 period. Comparing Fig. 5.4 with Fig. 5.2B, it is immediate to notice that all such events (A–D) were in synchrony with both dips or stagnation of per capita TEC, and steep falls of oil price.

5.4 A NATURAL OIL PRICE?

Fig. 5.5 shows the monthly Brent oil prices from January 2002 through June 2016, encompassing events "C" (monthly peak at $146/barrel of July 2008) and "D" (monthly peaks around $127/barrel on April 2011 and March 2012).

In each of those events, the residence time of oil prices above the $100/barrel and $80/barrel thresholds are very similar, while during the steep falls following a price spike, the threshold of $80/barrel is crossed very quickly, until a bottom price somewhat lower than $60/barrel, around the price levels of the 1990s, is reached.

Here we advance the hypothesis that an underlying force originated in the global interconnected community (or, if preferred, Smith's invisible hand) works to steer down the oil prices to levels well lower than $60/barrel

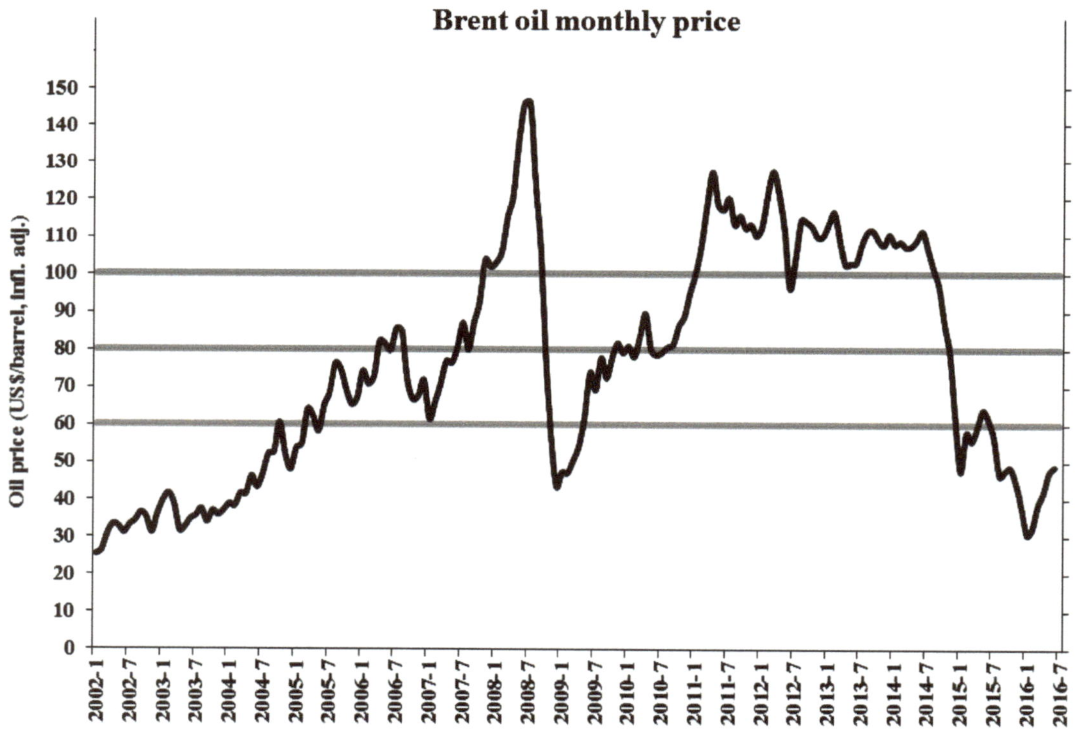

FIGURE 5.5

Series of monthly average inflation-adjusted Brent oil price, (January 2002—June 2016). *Reproduced from reference F. Meneguzzo, R. Ciriminna, L. Albanese, M. Pagliaro, The energy-population conundrum and its possible solution, arXiv:1610.07298 [physics.soc-ph], with kind permission.*

(in units of 2013 US dollar currency), in order to ensure the stability of the GDP "natural" growth rate, as derived by Eq. 5.2.

Supporting this hypothesis, in 2014 Murphy was noting that the average oil price during periods of economic growth over the past 40 years was under $40/barrel, and the average price during economic recessions was under $60/barrel. The previous observations corroborate the idea that oil has played a key role to sustain the GDP natural growth rate. Furthermore the sustainability of additional debt in Fig. 5.3B requires that the oil fraction in the energy mix should climb again to approximately 40% from current 33%, while oil prices should not exceed a threshold located somewhat between $40/barrel and $50/barrel, or possibly even lower. Alternatively other universally applicable energy sources, provided they are similarly abundant, concentrated and cheap as oil was until some time ago, should replace petroleum.

5.5 CONFLICTING DYNAMICS

Fig. 5.6A shows the worldwide monthly crude oil production series in million barrels daily, thus excluding other products, such as liquids produced from coal and gas, Orimulsion, biofuels such as biodiesel and ethanol, as well as other hydrocarbons, contributing to the overall oil supply. Moreover, the global crude oil supply is partitioned into outputs from the four biggest producers (United States, Russia, Saudi Arabia, and China), and the rest of the world.

Crude oil production in United States has been especially relevant for almost 70% of the increase in worldwide crude oil supply in the period 2005–14. Indeed while in June–August 2011 the United States crude oil output was around 5.5 million barrels/day, in April 2015 it achieved an astonishing volume of 9.7 million barrels/day, namely very few hundred thousand barrels lower than the historical peak occurred in November 1970.

Assuming January 2009 as the start date of significant tight-oil extraction, and considering the monthly US crude oil output since September 1, 1992, the long-term trend of the United States nontight crude oil output (production P_{nt} in units of million barrels daily), before tight-oil exploitation began, can be represented as a linear decreasing function of time, with around 90% of variance explained by Eq. 5.3:

$$P_{nt} = (-0.119 \pm 0.002) \cdot t + (6.92 \pm 0.01) \tag{5.3}$$

where t is time in units of years starting with $t = 0$ on September 1, 1992. Assuming such trend has continued after 2008, a rough estimate for the tight-oil output can be obtained after subtracting the linear trend represented by Eq. 5.3 from the complete series.

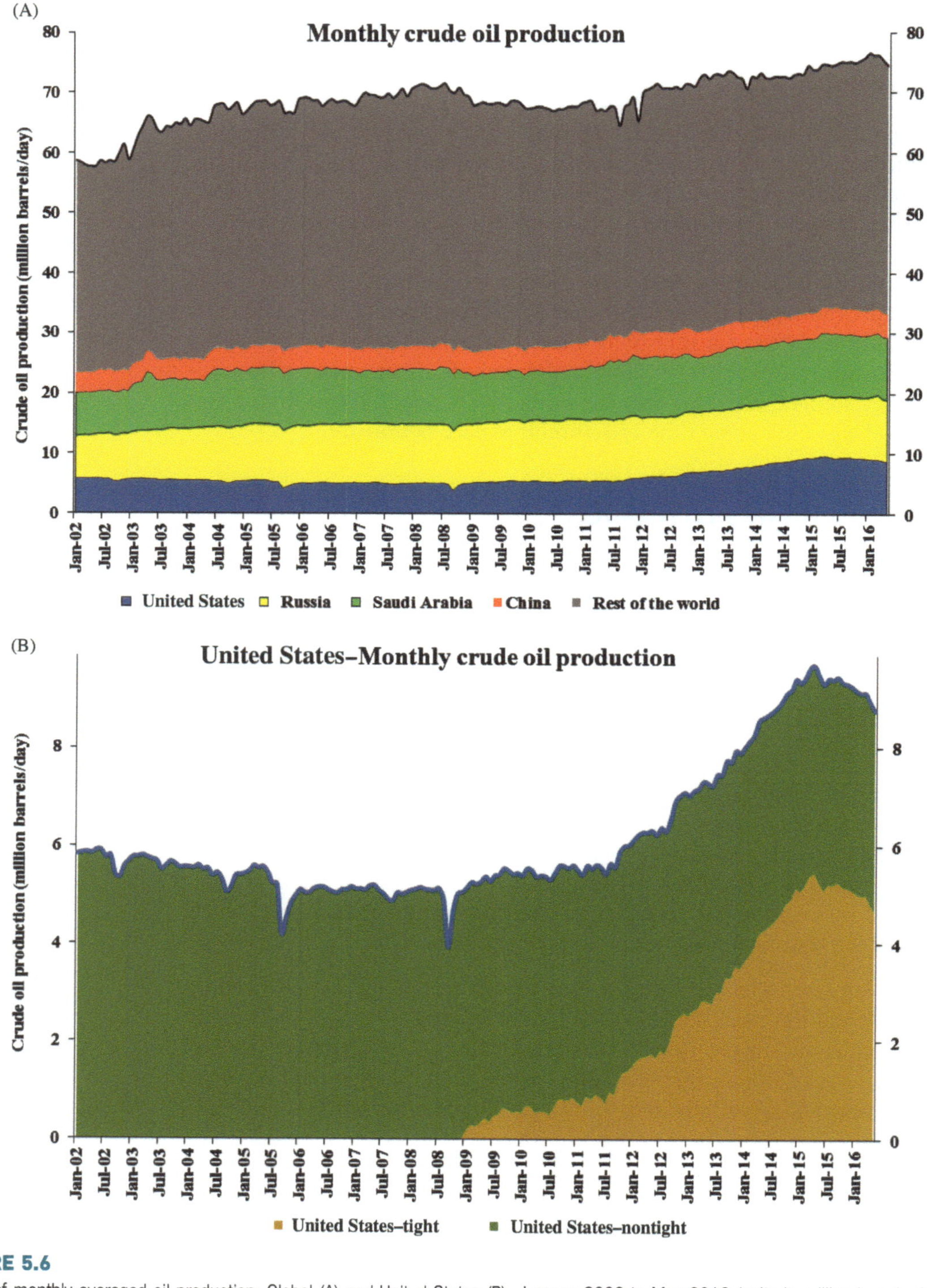

FIGURE 5.6

Series of monthly averaged oil production: Global (A) and United States (B), January 2002 to May 2016 (units in million barrels daily). *Reproduced from reference F. Meneguzzo, R. Ciriminna, L. Albanese, M. Pagliaro, The energy-population conundrum and its possible solution, arXiv:1610.07298 [physics.soc-ph], with kind permission.*

As mentioned earlier the tight economics of shale oil extraction indicates that roughly one-third of current production could be uneconomical at oil price around $60/barrel. Moreover, most tight and other unconventional oil resources extracted in the United States, Canada, Brazil, and Mexico would require prices close to $80/barrel [30] or even higher.

Indeed Fig. 5.6B displays that the exploitation of tight oil became significant in the overall US output during 2011, at the same time of the fast increase of oil prices persisting in the course of the sustained high prices of 2011–14. The unbalance between overall costs of US tight-oil production and prices of the oil commodity, which began falling few months earlier and crossed the $60/barrel on December 2014, as shown in Fig. 5.5, is self-evident.

In brief, unless conventional production from major producers is substantially increased, it is unlikely that unconventional resources such as tight oil, Canadian and Venezuelan oil sands, or biofuels, can help to keep the pace with the global oil need associated to the "natural" growth rate. Indeed Canadian oil sands are assessed uneconomical at oil prices even higher than those typical of US tight oil, while the heavily subsidized biofuels are losing momentum and substantially plateauing in output.

On the other hand, major crude oil producers, especially in the Middle East, have been lately extracting oil around their peak capacity despite the unfavorable market prices, thus shortening the time to peak and subsequent depletion [31].

A very recent model based on past production data for conventional and unconventional oil suggests that only certain Middle East countries might maintain their current production during the next decade [32], while all other producing regions will experience decline already between 2015 and 2020. As pessimistic as it may seem, that study omits to consider that even Middle East crude oil is not becoming more available outside the production area, mostly due to the fast growth of population and domestic consumption in those countries during the last 30 years.

5.6 THE FORTHCOMING SOLAR ECONOMY

If global population will keep on growing along the current trajectory, in 2016–25 about 800 million people will add to current population. Correspondingly, according to the linear relationship shown in Fig. 5.2A, in 2025 the TEC will have to be increased by about 1700 million tonnes of oil equivalent (MTOE) per year over the current level in order to feed the global wealth "natural" growth.

Even to keep the oil fraction in the energy mix at the current level around 33%, more than 11 million oil barrels per day should be added. On the other hand, in order to achieve the desirable threshold of 40% for the oil fraction in the energy mix, as much as additional 32 million oil barrels per day would be needed.

Such figure should be compared to the less than 10 million oil barrels per day added during the 2005–15 decade when the oil price was high (average around $90/barrel), including all liquids, out of which crude oil only accounted for 7.4 million barrels per day.

Recalling that most of the contribution to the increase in crude oil supply came from US tight oil, which is currently declining and likely unsustainable at the price levels compatible with global economy growth, perspectives could actually be even more uncertain.

Fig. 5.3A shows that natural gas and, even more, coal, have emerged as viable candidates to replace oil as major sources in the energy mix. About natural gas (leaving aside liquefaction meant to increase the concentration and ease transportability in the form of liquefied natural gas, consuming at least 25% of the internal energy) [33], it has limited concentration and requires the deployment of long pipelines such as those that connect Russia to Europe and, soon, to China. Though growing, the global production of natural gas is insufficient to compensate for the production decline of oil and coal.

As to coal, its basic problem is practical applicability only to power generation. The increase of its share in the energy mix has been chiefly due to impressive growth of lignite utilization in China (and to a lesser extent also in India). Yet in 2009 China became a net coal importer, with recent assessments placing the peak of lignite domestic production between 2025 and 2030 [34].

Uranium, as the source of nuclear power, enjoys by far the greatest energy concentration and its transportation is not an issue. Beyond very limited direct applications to the mobility sector such as military submarines, and icebreakers, its practical use for society lies in power generation, requiring a shift to electrification of energy end uses to achieve universal applicability. Although such shift is indeed possible and even desirable, relevant problem for uranium arise from availability and cost [35]. Production is forecast to peak in the second half of 2010s, followed by a slow decline up to 2025 and steeper afterwards, revealing insufficient even to feed existing and planned nuclear power plants [36].

A recent comprehensive study of Cartelle Barros and coworkers showed that the economics of the most representative RES such as high-temperature solar thermal (ST), onshore wind, solar PV, and small hydro power, are entirely comparable with that of oil, natural gas, and coal. Limited to solar PV, the earlier results are likely to derive from the already significant EROI of all its technological variants even in mid-latitude areas, and the steep energy learning curves of PV systems, making their deployment increasingly competitive with conventional sources at latitudes as high as 65 degrees [37].

Overall availability is not an issue at least for the solar and wind sources [38], but while deployment of high-temperature ST is profitable only at relatively low latitude, high insolation areas, solar PV can be deployed over a much wider portion of the global world [39]. As to the wind source, the geographical distribution of its availability is more dispersed than for solar PV.

Actually, the issues of availability, concentration and applicability show tight interlinks in the case of RES. Since their applicability is practically limited to power generation, the electricity transmissibility over long distances makes original concentration less than an issue, provided that availability is ensured around the clock and energy end uses can be fulfilled by electricity.

The share of RES over the TEC has grown over time. More in detail Fig. 5.7 shows that wind and solar PV (most of "other renewables") have been the main contributors to the recent strong acceleration experienced by the share of RES (from 7% in 2007 to 9.6% in 2015).

An analysis of the EROI dynamics unequivocally suggests declining EROI values for all fossil fuels [26], with the EROI of oil having likely halved in the short course of the first 15 years of the 21st century [40]. Consequently the overall increase in oil production needed to keep pace with the natural growth along the global population trajectory, that in the previous section was assessed up to 32 million oil barrels per day in 2025 over the current levels, could even be affected by substantial underestimation.

On the contrary the EROI of solar PV energy is experiencing a strong increase, due to more than double in 2020 in comparison to 2010 [37]. Consequently assessments of energy and capital investments needed to achieve the desirable replacement of the oil fraction in the energy mix, performed on the basis of past and even current data and performances, could be affected by a substantial overestimation.

A 100% grid penetration of intermittent wind-, water-, and solar-generated power for all purposes in the United States with electrification of virtually all energy end uses has been lately advocated by Jacobson and coworkers for the United States [41], and by us and Meneguzzo for Italy [42]. In such an electricity powered world based on intermittent renewable sources, a significant fraction of energy will be distributed across all battery vehicles [43].

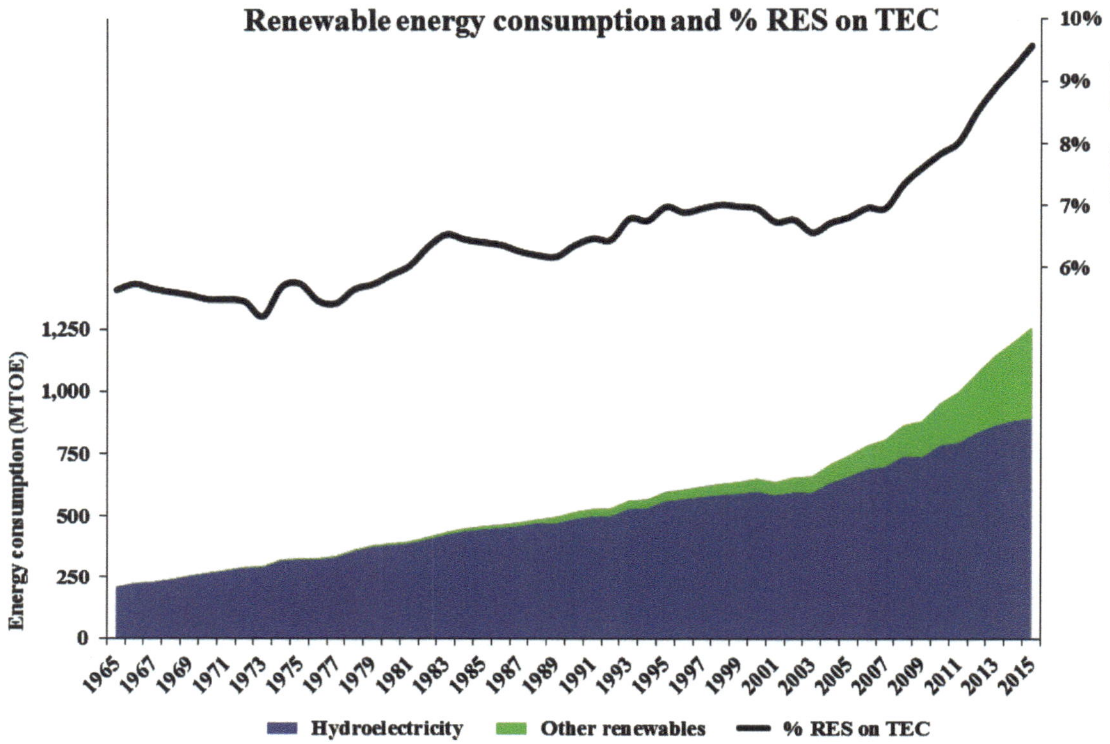

FIGURE 5.7

Renewable energy consumption, partitioned between hydroelectricity and other sources, and its percentage on total consumption, during 1965–2015. *Reproduced from reference F. Meneguzzo, R. Ciriminna, L. Albanese, M. Pagliaro, The energy-population conundrum and its possible solution, arXiv:1610.07298 [physics.soc-ph], with kind permission.*

While the intrinsic inertia of the complex energy-economy system works against any fast paradigm shift in energy generation and use, projecting replacement of oil with RES in the energy mix into several decades to come, evidence exists that locally much faster transitions have occurred [44], including experiences with RES in highly industrialized countries like Germany [45] and Italy [46], where the renewable electricity output in 2015 approached, respectively, 33% and 32.8% of domestic power demand.

Driven by low cost, quickly improving EROI, and broad social acceptability arising from their concomitant ecofriendly nature and positive impact of growing penetration on the prices formed in the wholesale electricity market [47,48], RES will continue their penetration in the energy mix of developed and developing countries at fast pace, until low cost energy storage technology such as Zhang's enzymatic fixation of hydrogen and CO_2 in renewable carbohydrates of high energy density will become widely available [49], making the solar economy a commonly shared reality.

5.7 BIOGLYCEROL-BASED POLYMERS AND CHEMICALS

Biopolymers and high value chemicals derived from glycerol are a safe bet for the future as chemical companies and investors evaluate their forthcoming production on significant scale. Facing conflicting demography and economic growth dynamics, oil currently selling at prices close to extraction costs, will shortly become a scarce resource. By 2025 the world will need 32 million barrels per day in addition to current 90 million barrels per day. This is simply not feasible, and will lead to permanent change in the energy and chemical industrial landscapes.

A new biochemical industry using both chemical and biochemical catalysis (biotechnology), which started to develop only when the price of oil surged between 2003 and 2008, will quickly develop across the world.

Glycerol will be one of the key platform chemicals of this new industry as it offers a key competitive advantage: Its unique chemical versatility due to uniquely high functionalization of the C3-backbone with three hydroxyl groups. It is this versatility, indeed, that made possible the synthesis of the first thermoset polymer, Plantics-GX, or that of squalane and vitamin D [50], or long chain C16-C18 fatty acids from glycerin and lignocellulosic hydrolysates [51], via fermentation over oleaginous yeasts, or even anticancer drug Taxol via cytochrome P450 hosted in *Escherichia coli* [52].

These are pricey molecules for which the availability of an efficient biotechnology process will immediately lead to practical application, as shown by the company that manufactures squalene and vitamin D in the United States.

Squalane, the best emollient available to the cosmetic industry, originates a \$200 million market growing at >9% yearly rate, largely dominated by scarcity and unstable supply as most squalene in its turn derives from shark liver or olive oil [53]. The biotechnology squalane derived from pure glycerol is of high purity (>99%) and completely devoid of color and odor, requiring only a better hydrogenation catalyst (compared to Ni Raney and Pd/C) for the complete hydrogenation of squalene under mild conditions [54].

The new biochemical industry will provide the solution to serious environmental problems such as, e.g., ocean littering due to oil-derived plastics. Truly biodegradable thermoplastic and thermoset polymers will replace current polypropylene, polyethylene, polyethylene terephthalate, polyesters, and acrylics at fast rate, regardless of the conservative nature of the petrochemical industry, and of the plastic industry in particular.

The glycerol-derived chemicals and biopolymers such as epichlorohydrin, mono-propylene glycol, PDO, PHA, glycerol polyesters, and the numerous other derivatives discussed in this book, whose production resisted 3 years of low oil prices, will be remembered as the first bioproducts of a long series.

Ever more, glycerol will be obtained by oleochemicals manufacturing from vegetable oil and tallow, while biodiesel production will continue to grow to eventually reach a *plateau*, and cultivation of drought resistant crops such as *Jatropha curcas* in arid lands, not in competition with food or feed, will provide another source of valued vegetable oil feedstock.

Eventually, mankind will continue to benefit in new and perhaps unexpected ways from what Milton Lesser in 1949 called "man's most versatile chemical servant": glycerin [55].

5.8 GLYCEROL: THE LIGNOCELLULOSIC BIOREFINERY ENABLER

Composed of cellulose, hemicellulose, and lignin, lignocellulosic biomass is an overabundant (>2 billion tonnes per year only from agricultural residues) [56] and intrinsically sustainable platform for the production of bio-based chemicals and polymers from which over 200 highly valued compounds can be derived via different derivatization methods, replacing the corresponding petroleum-based chemicals [57]. This will happen as long as the process to obtain the sugar feedstocks will be more economically viable, and thus profitable, than the competing oil-based routes starting from alkenes.

Leaf Resources Ltd., a small company based in Australia and listed on the Australia Stock Exchange (ASX:LER), has developed and is currently commercializing a process (trademarked Glycell) that uses crude biodiesel glycerol as a process enhancer to efficiently decompose plant biomass into lignin, cellulose, and hemicellulose at low temperature and pressure (Scheme 5.1) [58].

In detail the lignocellulosic biomass is first impregnated with sulfuric acid and steam at atmospheric pressure for 10–20 min and then fed into a reactor where crude glycerol is added. Sulfuric acid selectively depolymerizes and dissolves the hemicellulose [59]. After holding the mixture in the reactor for 25–30 min at 160°C a solid residue rich in cellulose and lignin is separated from a liquid fraction rich in C5 sugars, which also contains glycerol, some dissolved lignin and aqueous acid.

The solid fraction is washed with water and treated with hydrolyzing cellulase enzymes, which digest the cellulose and liberate monomeric glucose at up to 95% yield [60]. Rather than producing a combined hemicellulose and cellulose sugars product stream, the overall process affords two separate clean

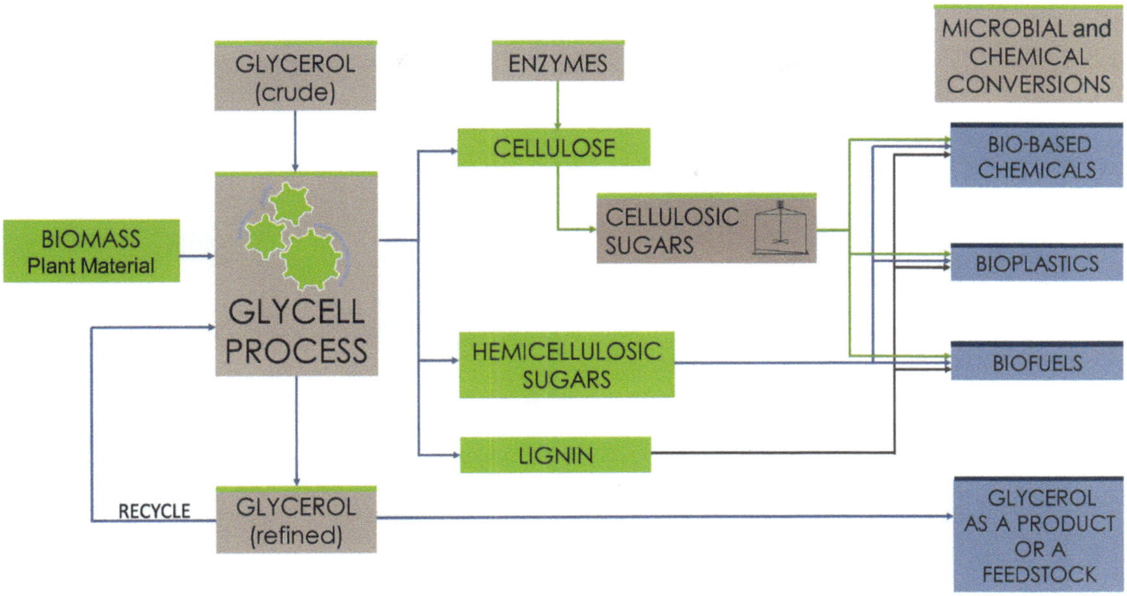

SCHEME 5.1

The biorefinery approach processing biomass into marketable products developed by Leaf Resources encompasses a key pretreatment step, the glycerol-based *Glycell* process. *Image courtesy of Leaf Resources.*

sugar product streams (viz., separate glucose rich and a xylose rich streams), a high-quality lignin, and glycerol as coproducts. The glycerol can be sold as a refined product or recycled.

A key advantage is that the Glycell process can use different biomass sources (Fig. 5.8), both wood-based (hard and softwood) and nonwood-based agricultural waste (bagasse, wheat straw, rice husk, palm oil waste, etc.). The process has been successfully tested on 4–6 dry tonnes biomass per day production rate at a facility in Ohio since November 2013.

Historically converting lignocellulosic biomass to sugars has been challenging, requiring harsh chemical treatment. The Germans at the end of World War II started to use concentrated hydrochloric acid for the saccharification of wood on industrial scale (the Scholler process), due to oil blockade [61]. The process produces large amounts of aqueous acid waste and was abandoned after the war. Other processes have been developed, such as dilute acid (Poet DSM); steam explosion (Beta renewable, Abengoa); ammonium fiber explosion (Dupont); supercritical water (Renmatrix); and concentrated acid (Virdia; company purchased by Stora Enso, and Ethanol Technologies in Australia). These new iterations of concentrated acid processes use simulated

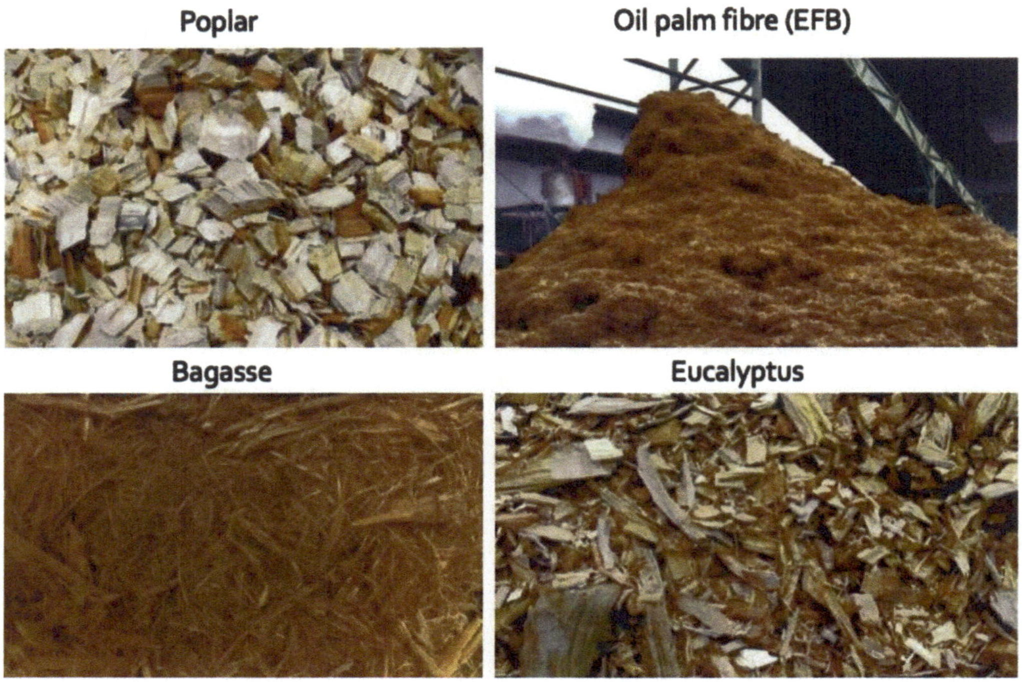

FIGURE 5.8
The Glycell process is versatile, enabling the use of wood-based and agricultural waste—based biomass. *Image courtesy of Dr. Les A. Edye, Leaf Resources.*

moving bed (SMB) chromatography to obtain a mixed sugars product and recover acid for reuse.

The overall Glycell process runs with high productivity and efficiency affording high yields of cellulose, with remarkable separation of hemicellulose and cellulose sugars, and eventually faster and more quantitative enzymatic hydrolysis of the cellulose fraction. For example, using sugarcane waste (bagasse) as feedstock, recovery of cellulose and hemicellulose is almost complete (cellulose yield of 94%) with quantitative (99%) conversion of cellulose to glucose syrup in 6 h only [2].

A model developed by the company shows that the process can produce cellulosic sugars at under $50/tonne when coproduct revenue is included and glycerol is not recycled. This compares with $220/tonne for sugars produced from the conversion of corn starch, the cheapest alternative and $280/tonne for raw sugar [62].

The Glycell process uses SMB chromatography to separate glycerol and the C5 sugars from other solutes in the liquid fraction. Testing using a SMB

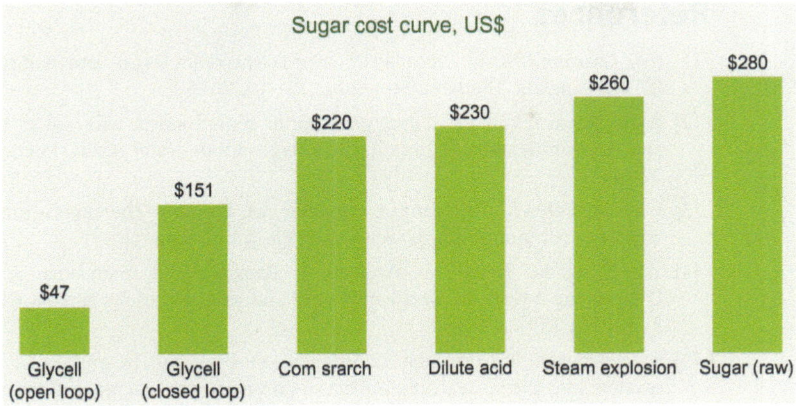

FIGURE 5.9
Sugar production costs. *Image courtesy of Leaf Resources.*

chromatography system on about 1 tonne of filtrate over 4 weeks confirmed that >95% of the glycerol was recoverable at 95% purity [58]. The latter grade can be easily brought to 99.7% and sold at $500/tonne. The process becomes open-loop, with glycerol sold replaced by 80% biodiesel glycerol purchased at $200/tonne, bringing the production cost of cellulosic sugars from $150/tonne to $47/tonne (Fig. 5.9).

Aiming at commercializing the technology, rather than entering into agreement with a petrochemical company, Leaf in early 2016 partnered with a biofuel company in the United States and using local knowledge and contacts provided by the new partner quickly signed agreements with Malaysian public innovation and bioeconomy agencies establishing close cooperation for the construction and operation of the first manufacturing facility using the Glycell process.

As discussed extensively throughout this chapter Malaysia, like numerous other southeast Asian and south American countries, has a keen interest in creating a successful bio-based industry going beyond the simple manufacturing of palm or coconut oil.

The main issue identified by the investment intelligence firm which analyzed the Glycell technology and its business perspectives in late 2015 is availability and supply logistics of both the glycerol and lignocellulosic biomass feedstocks and associated transport costs. Indeed, Malaysia would offer proximity to large amounts of bioglycerol and biomass available at low cost.

Assuming a biomass feedstock cost of $70/tonne, and a price of typical high value added chemicals obtained from cellulosic sugars at $1500/tonne, the industrial application of the Glycell process would translate into an after tax internal rate of return of 81% [62].

References

[1] R.F. Cascone, World Bio plastics market outlook: Status and issues, InnoBioPlast2016, BITEC, Bangkok, Thailand, September 15–16, 2016.

[2] Y.-H. Percival Zhang, Next generation biorefineries will solve the food, biofuels, and environmental trilemma in the energy–food–water nexus, Energy Sci. Eng. 1 (2013) 27–41.

[3] F. Meneguzzo, R. Ciriminna, L. Albanese, M. Pagliaro, The energy-population conundrum and its possible solution, arXiv:1610.07298 [physics.soc-ph].

[4] Y.Y. Deng, M. Haigh, W. Pouwels, L. Ramaekers, R. Brandsma, S. Schimschar, et al., Quantifying a realistic, worldwide wind and solar electricity supply, Glob. Environ. Chang 31 (2015) 239–252.

[5] F. Meneguzzo, R. Ciriminna, L. Albanese, M. Pagliaro, The great solar boom: a global perspective into the far reaching impact of an unexpected energy revolution, Energy Sci. Eng. 3 (2015) 499–509.

[6] I.C. Scotchman, Exploration for unconventional hydrocarbons: shale gas and shale oil, in: R. Hester, R. Harrison (Eds.), Fracking, Royal Society of Chemistry, Cambridge, 2014, pp. 69–103.

[7] S.H. Mohr, J. Wang, G. Ellem, J. Ward, D. Giurco, Projection of world fossil fuels by country, Fuel 141 (2015) 120–135.

[8] U. Bardi, Peak oil: the four stages of a new idea, Energy 34 (2009) 323–326.

[9] EIA, Drilling Productivity Report—April 2015, available at <http://www.eia.gov/petroleum/drilling/#tabs-summary-2>, 2015.

[10] EIA, Petroleum & Other Liquids, (2016), <http://www.eia.gov/dnav/pet/pet_sum_sndw_dcus_nus_w.htm>.

[11] Baker Hughes, North America Rig Count, <http://phx.corporate-ir.net/phoenix.zhtml?c=79687&p=irol-reportsother>.

[12] L. Doan, D. Murtaugh, Shale Oil Boom Could End in May After Price Collapse, bloomberg.com, April 13, 2015.

[13] C.J. Cleveland, P. O'Connor, Energy return on investment (EROI) of oil shale, Sustainability 3 (2011) 2307–2322.

[14] D.J. Murphy, The implications of the declining energy return on investment of oil production, Philos. Trans. R. Soc. A 372 (2014) 20130126.

[15] J.D. Hughes, Energy: a reality check on the shale revolution, Nature 494 (2013) 307–308.

[16] L.M.A. Bettencourt, J. Lobo, D. Helbing, C. Kühnert, G.B. West, Growth, innovation, scaling, and the pace of life in cities, Proc. Natl. Acad. Sci. USA 104 (2007) 7301–7306.

[17] United Nations, World population prospects, <https://esa.un.org/unpd/wpp/Download/Standard/Population/> (accessed 01.08.16).

[18] World Bank, World GDP (current US$), (2016). <http://data.worldbank.org/indicator/NY.GDP.MKTP.CD> (accessed 01.08.16).

[19] British Petroleum, BP Statistical Review of World Energy 2016, London, 2016.

[20] Joint Organisations Data Initiative (JODI), The JODI Oil World Database, <https://www.jodidata.org/oil/> (accessed 14.07.16).

[21] M. McLeay, A. Radia, R. Thomas, Money creation in the modern economy, Bank Engl. Q. Bull Q1 (2014) 14–27.

[22] G.E. Tverberg, Oil supply limits and the continuing financial crisis, Energy 37 (2012) 27–34.

[23] G.E. Tverberg, Financial issues affecting energy security, in: B. Sudhakara Reddy, S. Ulgiati (Eds.), Energy Security and Development: The Global Context and Indian Perspectives, Springer, New Delhi, India, 2015, pp. 69–83.

[24] G.E. Tverberg, Overly Simple Energy-Economy Models Give Misleading Answers, ourfiniteworld.com, 2016 (accessed 01.08.16).

[25] G.E. Tverberg, Energy limits: why we see rising wealth disparity and low prices, Our Finite World, ourfiniteworld.com, 2016 (accessed 01.08.16).

[26] C.A.S. Hall, J.G. Lambert, S.B. Balogh, EROI of different fuels and the implications for society, Energy Policy 64 (2014) 141–152.

[27] I. Chapman, The end of peak oil? Why this topic is still relevant despite recent denials, Energy Policy 64 (2014) 93–101.

[28] J.W. Murray, J. Hansen, Peak oil and energy independence: myth and reality, Eos Trans, 94, AGU, 2013, pp. 245–246.

[29] G.E. Tverberg, Overview of our energy modeling problem, Our Finite World, http://ourfiniteworld.com, 2015 (accessed 01.08.16).

[30] Bank of Canada, Monetary Policy Report, Ottawa, 2015.

[31] S.H. Hanke, On the Price of Oil, Global Asia, June 2016.

[32] M. Dittmar, Regional Oil Extraction and Consumption: A simple Production Model for the Next 35 Years Part I, 2016. arXiv:1601.07716 [physics.soc-ph].

[33] A. Vatani, M. Mehrpooya, A. Palizdar, Energy and exergy analyses of five conventional liquefied natural gas processes, Int. J. Energy Res. 38 (2014) 1843–1863.

[34] J. Wang, L. Feng, G.E. Tverberg, An analysis of China's coal supply and its impact on China's future economic growth, Energy Policy 57 (2013) 542–551.

[35] J.J. Cartelle Barros, M. Lara Coira, M.P. de la Cruz López, A. del Caño Gochi, Assessing the global sustainability of different electricity generation systems, Energy 89 (2015) 473–489.

[36] M. Dittmar, The end of cheap uranium, Sci. Total Environ. 461 (2013) 792–798.

[37] M. Görig, C. Breyer, Energy learning curves of PV systems, Environ. Prog. Sustain. Energy 35 (2016) 914–923.

[38] N. Armaroli, V. Balzani, Solar electricity and solar fuels: status and perspectives in the context of the energy transition, Chem. Eur. J. 22 (2016) 32–57.

[39] J. Hernández-Moro, J.M. Martínez-Duart, Analytical model for solar PV and CSP electricity costs: present LCOE values and their future evolution, Renew. Sustain. Energy Rev. 20 (2013) 119–132.

[40] S. Sgouridis, U. Bardi, D. Csala, A Net Energy-based Analysis for a Climate-constrained Sustainable Energy Transition, 2015. arXiv:1503.06832 [physics.soc-ph].

[41] M.Z. Jacobson, M.A. Delucchi, M.A. Cameron, B.A. Frew, Low-cost solution to the grid reliability problem with 100% penetration of intermittent wind, water, and solar for all purposes, *Proc. Natl. Acad. Sci.* USA, 112, 2015, pp. 15060–15065.

[42] F. Meneguzzo, R. Ciriminna, L. Albanese, M. Pagliaro, Italy 100% Renewable: A Suitable Energy Transition Roadmap, arXiv:1609.08380 [physics.soc-ph].

[43] L. Albanese, R. Ciriminna, F. Meneguzzo, M. Pagliaro, The impact of electric vehicles on the power market, Energy Sci. Eng. 3 (2015) 300–309.

[44] B.K. Sovacool, How long will it take? Conceptualizing the temporal dynamics of energy transitions, Energy Res. Soc. Sci 13 (2016) 202–215.

[45] C. Morris, Germany is 20 years away from 100 percent renewable power – not!, energytransition.de, January 5, 2016.

[46] GSE, Energia da fonti rinnovabili in Italia. Dati preliminari 2015, gse.it, February 29, 2016.

[47] M. Pagliaro, F. Meneguzzo, F. Zabini, R. Ciriminna, Assessment of the minimum value of photovoltaic electricity in Italy, Energy Sci. Eng. 2 (2014) 94–105.

[48] F. Meneguzzo, R. Ciriminna, L. Albanese, M. Pagliaro, The remarkable impact of renewable energy generation in Sicily onto electricity price formation in Italy, Energy Sci. Eng. 4 (2016) 194–204.

[49] Y.-H.P. Zhang, W.D. Huang, Constructing the electricity-carbohydrate-hydrogen cycle for sustainability revolution, Trends Biotechnol. 30 (2012) 301–306.

[50] S. O'Connor, Challenges and opportunities in commercializing a fermentation based production platform, World Congress on Industrial Biotechnology, April 20, 2016.

[51] J.V. Álvarez, MicroBioOil, World Bio Markets, Amsterdam, March 14–16, 2016.

[52] Br. W. Biggs, C.G. Lim, K. Sagliani, S. Shankar, G. Stephanopoulos, M. De Mey, et al., Overcoming heterologous protein interdependency to optimize P450-mediated taxol precursor synthesis in *Escherichia coli*, Proc. Natl. Acad. Sci. USA 113 (2016) 3209–3214.

[53] Global Market Insights, Squalene Market Size by Source (Shark Liver, Vegetable, Synthetic), by Application (Cosmetics, Supplements, Pharmaceuticals), Industry Analysis Report, Regional Outlook, Application Potential, Price Trends, Competitive Market Share & Forecast, 2015–2022, Ocean View, Delaware, April 2016.

[54] V. Pandarus, R. Ciriminna, S. Kaliaguine, F. Béland, M. Pagliaro, Heterogeneously catalyzed hydrogenation of squalene to squalane under mild conditions, ChemCatChem 7 (2015) 2071–2076.

[55] M.A. Lesser, Glycerin—Man's most versatile chemical servant, J. Chem. Educ. 26 (1949) 327–328.

[56] Y.Y. Tye, K.T. Lee, W.N.W. Abdullah, C.P. Leh, The world availability of non-wood lignocellulosic biomass for the production of cellulosic ethanol and potential pretreatments for the enhancement of enzymatic saccharification, Renew. Sustain. Energy Rev. 60 (2016) 155–172.

[57] F.H. Isikgor, C. Remzi Becer, Lignocellulosic biomass: a sustainable platform for production of bio-based chemicals and polymers, Polym. Chem. 6 (2015) 4497–4559.

[58] L. Edye, A. Baker, M. Sabourin, Glycell—Leaf Resources' pretreatment process for the conversion of lignocellulosic biomass to fuels and chemicals, 37th Symposium on Biotechnology for Fuels and Chemicals, San Diego, April 27–30, 2015.

[59] Edison Investment Research, Leaf Resources, Game-changer technology for bio-based products, Sydney, Australia, September 2015.

[60] X. Wu, J. McLaren, R. Madi, D. Wang, Biofuels from lignocellulosic biomass, in: O.V. Singh, S.P. Harvey (Eds.), Sustainable Biotechnology, Springer, New York, 2009, p. 19.

[61] W.L. Faith, Development of the scholler process in the United States, Ind. Eng. Chem. 37 (1945) 9–11.

[62] Leaf Resources, Investor Presentations, asx.com.au, August 8, 2016.

Index

Note: Page numbers followed by "*f*" and "*t*" refer to figures and tables, respectively.

C

D

E

F

G

Kolibri 360 GmbH
Ferdinand-Jühlke-Straße 7
99095 Erfurt, Deutschland
produktsicherheit@kolibri360.de